D1522628

# Chemotherapy of Tropical Diseases

Critical Reports on Applied Chemistry Volume 21

# Chemotherapy of Tropical Diseases

edited by M. Hooper



Published for the Society of Chemical Industry by
John Wiley & Sons
Chichester · New York · Brisbane · Toronto · Singapore

***Library of Congress Cataloging-in-Publication Data:***
Chemotherapy of tropical diseases.
(Critical reports on applied chemistry; v. 21)
Includes bibliographies and index.
1. Tropical medicine. 2. Anti-infective agents.
3. Parasitic diseases — Chemotherapy. I. Hooper, M.
II. Title. III. Series. [DNLM: 1. Leishmaniasis — drug therapy. 2. Mycobacterium Infections — drug therapy. 3. Nematode Infections — drug therapy.
4. Tropical Medicine. 5. Trypanosomiasis — drug therapy. WC 680 C517]
RC961.C43 1987 616.9′6′00913 86-16005

ISBN 0 471 91241 7

***British Library Cataloguing in Publication Data:***
Chemotherapy of tropical diseases: the problem and the challenge. — (Critical reports on applied chemistry; v.21)
1. Chemotherapy — Tropics
I. Hooper, M. II. Society of Chemical Industry III. Series
616.9′88′3061 RM263

ISBN 0 471 91241 7

Typeset by Activity Ltd., Salisbury, Wiltshire.
Printed and bound in Great Britain

# Contents

# List of Contributors

**J. Barrett** *Department of Zoology, University of Wales, Aberystwyth, Dyfed SY23 3DA*

**J.R. Brown** *Tropical Diseases Chemotherapy Research Unit and Department of Pharmaceutical Chemistry, Sunderland Polytechnic, Sunderland SR2 7EE*

**D.A. Denham** *London School of Hygiene and Tropical Medicine, Keppel Street, London WC1E 7HT*

**L.G. Goodwin** *The Wellcome Trust, 1 Park Square West, London NW1 4LJ*

**M. Hooper** *Tropical Diseases Chemotherapy Research Unit and Department of Pharmaceutical Chemistry, Sunderland Polytechnic, Sunderland SR2 7EE*

**D.E. Minnikin** *Department of Organic Chemistry, The University, Newcastle upon Tyne NE1 7RU*

# Editor's introduction

Tropical diseases and their treatment are currently the subject of widespread debate, which is often fierce and sometimes acrimonious (see, for example, Melrose, 1982; Taylor, 1982). One writer (Cerami, 1979) regards some tropical diseases as orphan diseases. He asserts that the huge costs of research prevent the development of new drugs. Developing countries which already spend very high proportions of their total health expenditure on drugs (Lionel, 1981) are unable to contribute to the costs of essential research programmes. Other writers have demonstrated the very real concern of the pharmaceutical industry (O'Driscoll, 1983), which provides some drugs at fair prices to developing countries and has developed some new drugs and vaccines for specific diseases (Taylor, 1982; Jones, 1983).

In the midst of this debate this volume seeks to present both information and challenge in the area of drug development and tropical disease. The terminology of tropical diseases is a common source of confusion and even despair! The diseases mentioned herein are all parasitic diseases which are unfamiliar to most scientists in the Western world. *Mycobacterium leprae*, a prokaryotic organism*, is the causative agent of leprosy. It is closely related to the better known *Mycobacterium tuberculosis*. Alternative names for leprosy are Hansen's disease or hanseniasis. Then follow the eukaryotic* protozoa in order of structural complexity. These are essentially single cell organisms and include plasmodia, some species of which cause malaria, and trypanosomes, flagellate protozoa which include the causative agents of trypanosomiasis, the African form is known as sleeping sickness and the South American disease as Chaga's disease, and leishmaniasis.

Worm infections in man are very widespread. The principal examples referred to in this volume are filariases caused by filarial worms which are nematodes (threadworms). The major clinical manifestations are enormous swellings (elephantiasis) or blindness (river blindness). Trematodes are flatworms, sometimes referred to as flukes, which include the blood flukes, of the genus *Schistosoma*, responsible for human schistosomiasis (bilharzia). With the exception of *M. leprae* these parasites have complex life cycles involving, in addition to man, a secondary host (mosquitoes, various flies, bugs, snails). The four chapters in this book provide overwhelming evidence of the need for new drugs and drug delivery systems for the treatment of leprosy, filariasis and trypansomiasis — diseases which affect millions of people. Two of the agents of these diseases are part of larger groups of parasites, filarial worms and trypansomes, which also cause widespread diseases in animals leading to still further enormous social and economic loss.

The chemotherapy of any disease requires interdisciplinary groups of different specialists each making their own distinctive contribution. This is reflected in the authors of this volume who represent biology, chemistry, biochemistry and medicinal chemistry. Chapter 1 is an authoritative survey of the major tropical diseases currently forming the World Health Organization's Special Programme on Tropical Diseases. Chapter 2 looks at the detailed chemistry and biochemistry of the mycobacterial cell wall which provides a unique target for new drug development. Chapter 3 describes the biology, biochemistry and chemotherapy of human filariasis and Chapter 4 the biochemistry and medicinal chemistry of trypanosomiasis and leishmaniasis. Each author has been both critical and creative in their review and has identified possible selective new targets for the design of novel chemotherapeutic agents.

This volume demonstrates the enormous scientific challenge of these longstanding human diseases, both in understanding the disease processes and the ways in which selective new medicines may be developed. They also draw attention to areas of contemporary research into new medicines for a variety of diseases in the developed world. For example, γ-aminobutyric acid (gaba) is a major inhibitory transmitter in the mammalian central nervous system. It is currently the subject of intense study. Surprisingly, the large molecules of ivermectin, a veterinary drug, appear to act by enhancing the effects of gaba in filariae (Campbell, 1985). If drug companies and academic departments would submit gaba agonists and antagonists for testing, new filaricidal drugs might be discovered.

α,α-Difluoromethylornithine, designed as an anticancer agent, is effective against trypanosomes (Fozard and Koch-Weser, 1982). Acquired immune deficiency syndrome (AIDS) is now threatening to develop into a worldwide disease. This viral disease cannot be effectively treated with drugs at present. A major lead is the recognition that suramin, an old filaricidal and trypanocidal drug, is a potent reverse transcriptase inhibitor of this retrovirus *in vitro* (Mitsuya *et al.*, 1984). Pentamidine, another old trypanocidal drug, is currently the preferred treatment for the AIDS-related *Pneumocystis carnii* pneumonia (Cole, 1984). Dapsone, the classical antileprotic drug, in addition to its use in dermatology, has also been reported to be effective in the treatment of Kaposi's sarcoma in a case of AIDS (Poulsen *et al.*, 1984).

All the parasitic diseases described in this volume are associated with unusual and complex immunological changes. The unravelling of these responses may well produce vital information which will assist in the understanding and treatment of other diseases involving the immune system, e.g. cancer and arthritis. Here, then, is food for thought and action which provides an opportunity for both scientific endeavour and service to humanity.

*A simple distinction is that prokaryotic cells contain DNA which is not organised into chromosomes or enclosed in a nuclear membrane whilst eukaryotic cells contain DNA enclosed in a nuclear membrane and usually other organelles eg. mitochondria.

It is with great pleasure that I commend this volume of *Critical Reports in Applied Chemistry* and express my thanks to the contributing authors.

M. HOOPER
*Tropical Diseases Chemotherapy Research Unit and Department of Pharmaceutical Chemistry, Sunderland Polytechnic, Sunderland SR2 7EE*

## References

1. Campbell, W.C. (1985). *Parasitology Today*, **1**(1), 10–16.
2. Cerami, A. (1979). *The Sciences*, 22–25.
3. Cole, H. (1984). *J. Am. Med. Ass.*, **252**(15), 1987–1988.
4. Fozard, J.R., and Koch-Weser, J. (1982). *Trend. Phar.*, **3**, 107–110.
5. Jones, T.M. (1983). *Pharm. J.*, 301–307.
6. Lionel, N.D.W. (1981). *Trend. Phar.*, **2**(5), I.
7. Melrose, D. (1982). *Bitter Pills: Medicines and the Third World Poor*, Oxfam, Oxford.
8. Mitsuya, H., Popovic, M., Yarchoan, R., Matsushita, S., Gallo, R.C., and Broder, S. (1984). *Science*, **226**, 172–174.
9. O'Driscoll, L. (1983). *Chem. Britain*, **19**(12), 984.
10. Poulsen, A., Hultberg, B., Thomsen, K., and Wantzin, G.L. (1984). *Lancet*, **i**, 560.
11. Taylor, D. (1982). *Medicines, Health and the Poor World*, Office of Health Economics, London.

# 1 Chemotherapy and tropical disease: The problem and the challenge

L.G. Goodwin
The Wellcome Trust, London



## 1.1 Introduction

Communicable parasitic disease continues to present a major problem in developing countries in the tropics. In many temperate countries improvements in hygiene and living standards have removed the threat of widespread endemic and epidemic infections, and debilitating, chronic disease has been controlled. In the tropics the wide distribution of invertebrate vectors, the impossibility of killing them all and the fact that for some diseases such as trypanosomiasis, leishmaniasis and filariasis, wild animals act as reservoir hosts for the parasites greatly adds to the problem.

The burden of parasitic disease in the tropics is immense; a recent estimate made by the World Health Organization (WHO) is shown in Table 1.1.

Improvements in hygiene, control of insect vectors and the provision of clean, piped water supplies would greatly reduce the risks, but these measures are expensive in money and manpower, and they take time. They do not solve the

**Table 1.1** Numbers of people exposed to and infected with tropical diseases

| Disease | Number exposed | Number infected |
|---|---|---|
| Malaria | 2000 million | 200 million |
| Trypanosomiasis | | |
| African | 50 million | 20 000 new cases a year |
| South American | 65 million | 20 million |
| Leishmaniasis | ? | 0.5 million new cases a year |
| Schistosomiasis | 600 million | 200 million |
| Filariasis | | |
| Onchocerciasis | 500 million | 30 million |
| Lymphatic | 900 million | 90 million |
| Leprosy | 1400 million | 11 million |

problem of people who already have disease and provide a continuing source of reinfection; improvement of the environment must go hand in hand with the effective treatment of patients.

Although the treatment of many diseases of the developed world has been revolutionized in recent years by the introduction of new, effective drugs, the parasitic diseases of the tropics have, on the whole, been sadly neglected.

There are good reasons for this. The research is difficult and unpredictable, partly because we lack sufficient knowledge of the parasites' metabolic processes and their pathological effects on their hosts. Nowadays, the cost of toxicology for the clearance of any promising drug for clinical trial is measured in millions of pounds and few pharmaceutical companies are willing to take risks. If an advance is made and a new drug is introduced, it is not unusual for it to be 'pirated' by an organization in a country that takes little notice of patent laws and puts the drug on the market at a much lower price — which it can afford to do, having had no research and development costs to meet. New drugs can never be very cheap and impoverished governments in developing countries understandably buy as much as they can with the limited funds available. The people who need treatment seldom have enough money to buy the medicines for themselves. There is therefore very little incentive for the pharmaceutical industry to risk much in the way of research work on medicines for the tropics; many companies have withdrawn all efforts in this direction. This is a wry paradox because the highly successful science of chemotherapy had its origin in Ehrlich's efforts to cure trypanosome infections with arsenic compounds.

It has sometimes been asserted that the industry has already provided a range of drugs for tropical parasitic diseases and that treatments should not be abandoned just because they are old. This argument has no real force for trypanosomiasis, leishmaniasis and filariasis because the compounds available suffer from unacceptable disadvantages that make them difficult, dangerous or time consuming to administer.

However, in the field of intestinal anthelmintics, the pharmaceutical industry, through its research on veterinary chemotherapy, has made an invaluable contribution. There is a better world market for drugs for treating animal parasitic disease than for human infections in the tropics, and drugs are now available that are often effective in a single dose. Several of these have been cleared for human use and are proving to be of great value for removing intestinal helminths. In addition, compounds effective against liver and blood flukes, and against the tissue cysts of cestodes, have come into human medicine through the veterinary field.

## 1.2 Needs

The people in need of treatment are, in the main, in impoverished rural and urban communities living in poor housing, with limited access to medical attention in countries in which basic health/care facilities have not yet been adequately developed. Great efforts are now being made by WHO in the 'Health for all by the year 2000' campaign to improve these facilities. Governments are being persuaded to invest their resources in modest but well-organized and well-sited medical centres, instead of sinking large sums in prestigious hospitals in large cities. Such hospitals are ruinously expensive to keep equipped, supplied and staffed, and few patients in these countries can afford either the money or the time for hospital treatment in town — they need to get back to work on the farm or in the factory and to support their families.

Drugs that have to be given repeatedly for more than a day or two are inappropriate; ideally an effective treatment should be accomplished in a single dose, especially if attempts are made to give mass treatment to populations in rural areas. Supplies of tablets handed to patients in the expectation that they will be taken at specified intervals are usually wasted; the tablets are often thrown away or sold or given to others. For mass treatment, a drug must be very safe, especially for children and pregnant women.

A further serious problem, especially with protozoal infections and leprosy, is the development of drug resistance. Strains of parasite exist, are selected or emerge during the use of a drug that resist its toxic action and render it useless. It is therefore important to avoid the misuse of a drug under conditions that favour the emergence of resistant parasites and to have available an alternative effective treatment.

## 1.3 Recent progress

The World Health Organization's Parasitic Diseases Programme has for many years assisted the efforts of field and laboratory research workers attempting to improve the diagnosis and treatment of tropical infections. Since 1977 the UNDP/World Bank/WHO Special Programme for Research and Training in Tropical Diseases has substantially added to the effort. The aim has been to interest scientists from many disciplines all over the world in the biology, immunology and chemotherapy of the six most prominent tropical diseases —

malaria, trypanosomiasis, leishmaniasis, schistosomiasis, filariasis and leprosy — and to strengthen the facilities and competence of research institutions in the countries where the diseases occur. The programme has gathered momentum very rapidly, partly because of recent exciting developments of techniques in the fields of immunology, genetics and molecular biology. Grants from the Special Programme enable scientists to carry out work that would have been impossible with existing commitments, budgets and staff. At the same time, basic accommodation, equipment, techniques, expertise and leadership have been provided by the institutions at which the work is done — an efficient and cost-effective operation. Considerable progress has been made in chemotherapy, for which the expertise and collaboration of the pharmaceutical industry has been essential. Collaborative projects for screening compounds and advancing some of them to clinical trial have already led to the discovery of several promising new drugs (UNDP/World Bank/WHO, 1984).

In addition, there has been the continuing support for more basic studies at universities and other institutions by government agencies in the developed countries and by charitable foundations. These foundations are now putting considerable sums towards the study of neglected diseases and progress is being made in understanding the biology of the parasites and their relationships with their hosts.

## 1.4 Existing treatment and promising new drugs

### *1.4.1 Malaria*

Malaria is still an important obstacle to health and progress; strenuous efforts to eradicate it have met with only limited success and the disease has returned to some of the regions from which it had once been cleared (Figure 1.1). Major factors in its return have been the development of resistance to insecticides in the mosquito and poor surveillance, often caused by population movements and political instability. Even more serious has been the spread of parasites resistant to the existing antimalarial drugs (Figure 1.2); there is an urgent need for new drugs with activity against the resistant strains.

Malaria has received more attention than other tropical parasitic infections (UNDP/World Bank/WHO, 1983; Peters and Richards, 1984; WHO, 1984a). It has always been a menace to non-immune soldiers deployed in endemic areas, and the Second World War, the Korean War and the Vietnam War all stimulated research into antimalarial drugs. Shortage of quinine during the Second World War led to the use of mepacrine and the 4-aminoquinoline derivatives chloroquine (**1**) and amodiaquine, and the introduction of the antifolate drugs proguanil (**2**) and pyrimethamine (**3**). The Korean War established the value of the 8-aminoquinoline compound primaquine (**4**) for terminating attacks of *Plasmodium vivax* malaria. The Vietnam War stimulated the screening of many thousands of substances at the Walter Reed Institute in the United States and gave

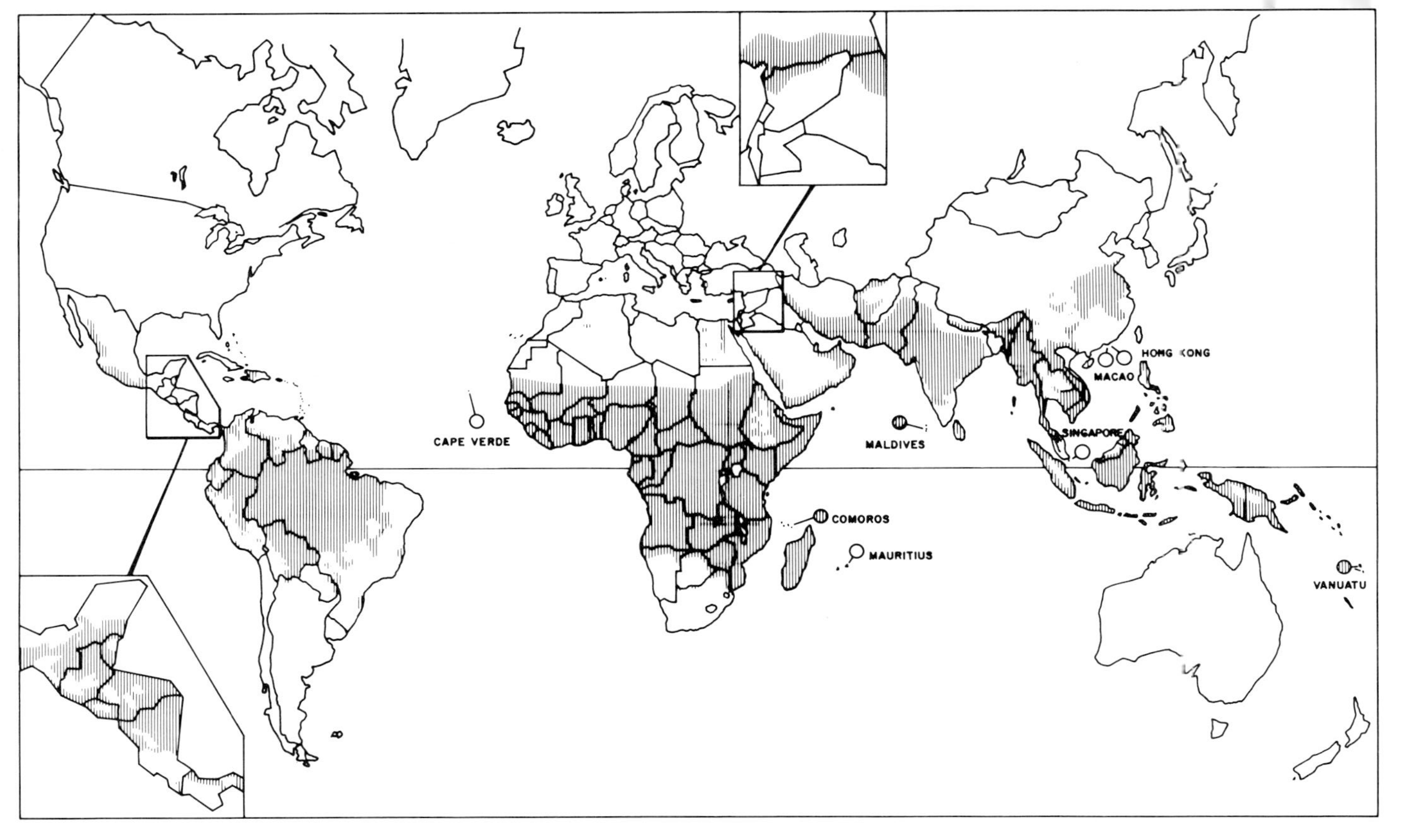


**Figure 1.1** Areas in which malaria is transmitted (1982). (*From* Vaccination Certificate Requirements for International Travel and Health Advice to Travellers 1984. *Reproduced by permission of World Health Organization, Geneva.*)

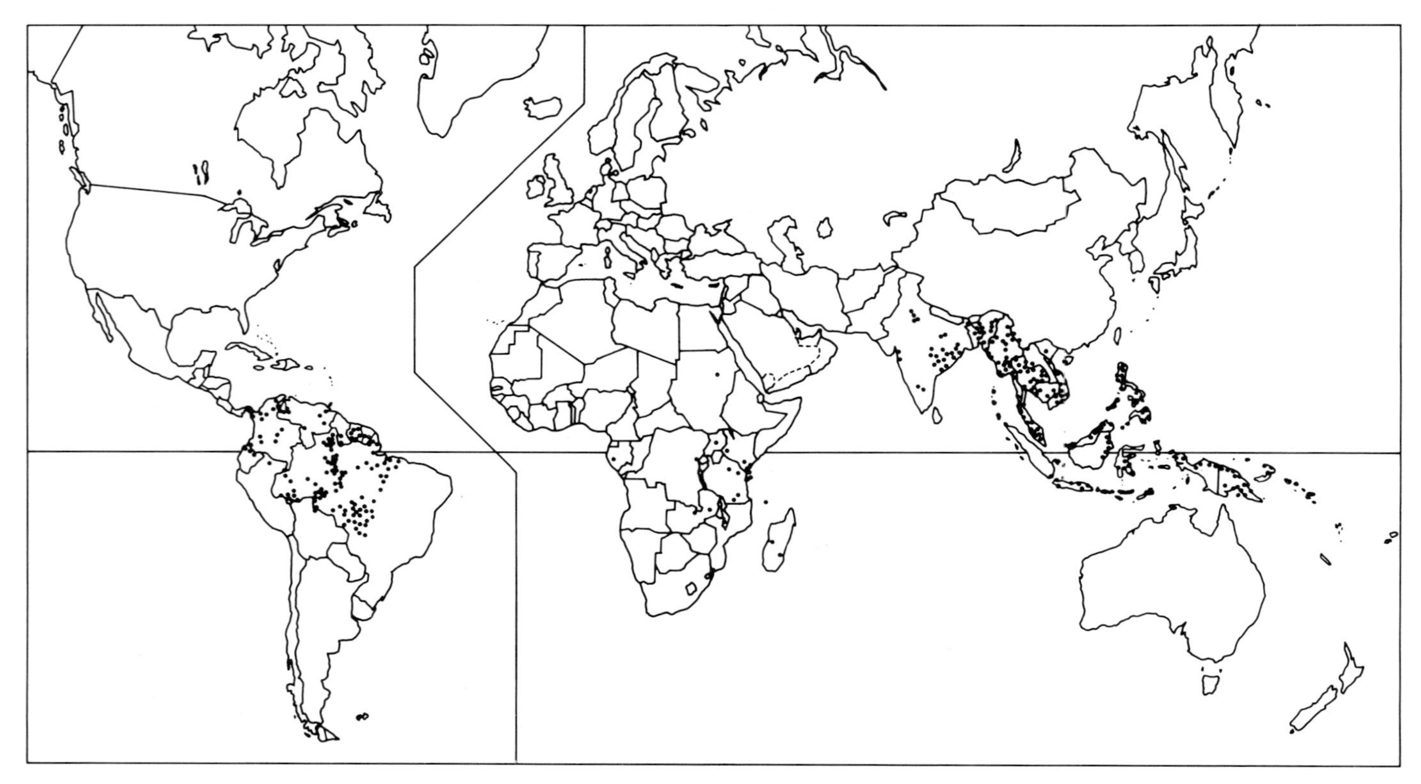

**Figure 1.2** Areas where chloroquine-resistant *Plasmodium falciparum* malaria has been reported. (*From* Vaccination Certificate Requirements for International Travel and Health Advice to Travellers 1984. *Reproduced by permission of World Health Organization, Geneva.*)

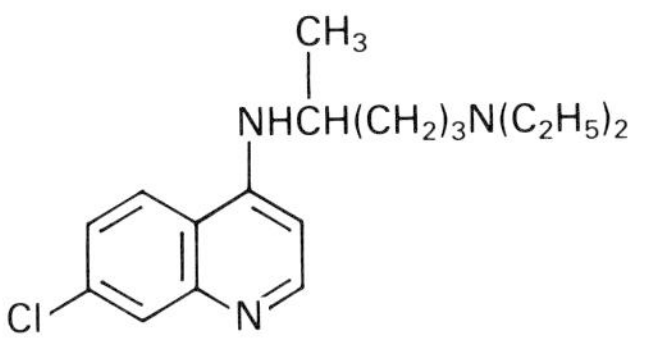

(**1**) Chloroquine

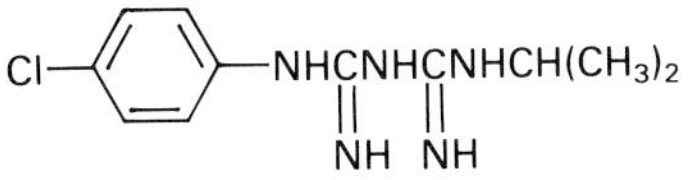

(**2**) Proguanil

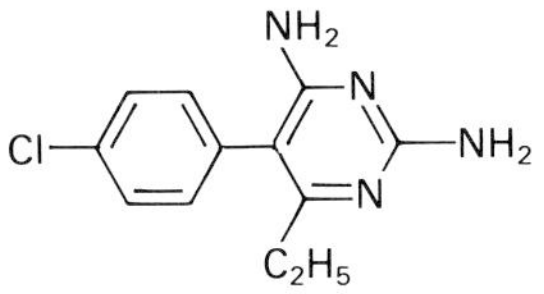

(**3**) Pyrimethamine

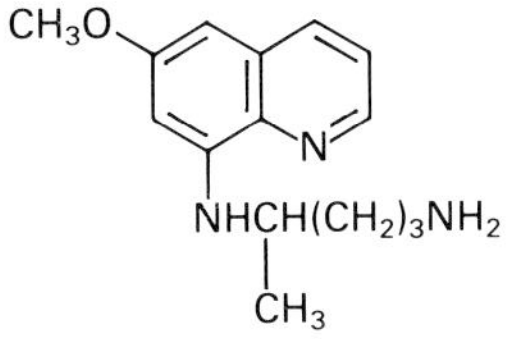

(**4**) Primaquine

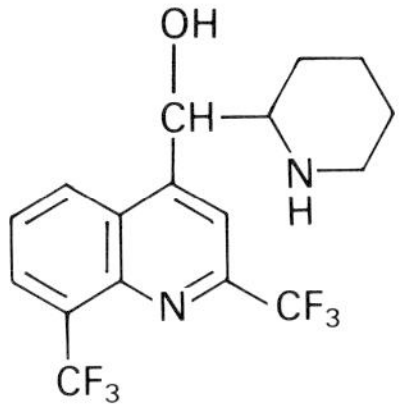

(**5**) Mefloquine

rise to the quinolinemethanol compound mefloquine (**5**) and other promising leads. Mefloquine is now under extensive clinical trial at three centres under WHO control.

One of the centres is in Thailand, where drug resistance to the 4-aminoquinolines, to potentiating mixtures of pyrimethamine and long-acting sulphonamides, and even to quinine itself, has become established. Figure 1.2 shows the recent distribution of chloroquine-resistant malaria. It is still spreading and regular assessment is necessary to track the changing areas of resistance; test kits are available from the WHO for determining the degree of resistance of the parasites *in vitro*.

*(a) Mefloquine* Mefloquine (**5**) has an action similar to that of quinine. It is effective against all species of malaria parasite and against strains resistant to chloroquine and the antifolate drugs. It cures *P. falciparum* malaria attacks in a single dose (Sweeney, 1981). However, parasites resistant to mefloquine have already been encountered during clinical trials and unless this new, very useful drug is used with great circumspection, it might quickly become useless in some areas.

In the treatment of tuberculosis and leprosy it has been found that the simultaneous administration of effective doses of several drugs with differing modes of action from the outset is less likely to give rise to resistance than the deployment of the drugs one at a time. There is evidence that this also applies to malaria and it is therefore intended (except in special cases) not to use mefloquine alone, but always in combination with pyrimethamine (**3**) and sulphadoxine in a determined effort to retain its efficacy for as long as possible. Mefloquine should never be used in areas where the other drugs are effective (WHO, 1984a).

The Walter Reed programme also selected other chemical series with antimalarial activity. The phenanthrenemethanol derivative, halofantrine (**6**)

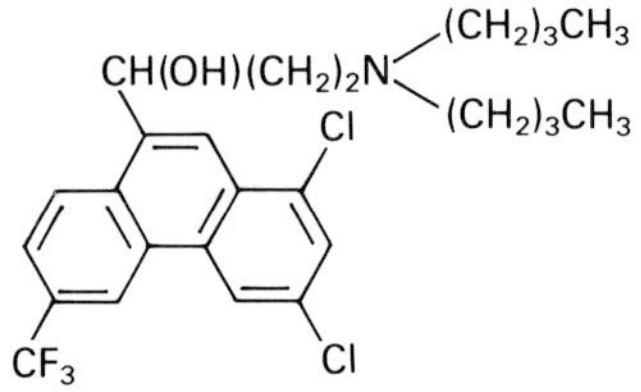

(**6**) Halofantrine

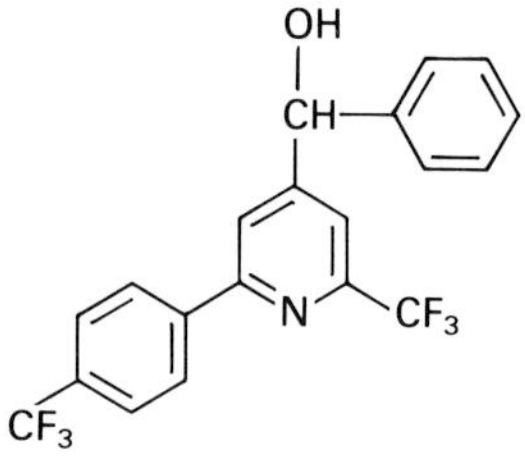

(**7**) WR 180409

| | R |
|---|---|
| (**8**) Quinghaosu (artemesinine) | O |
| (**9**) Artemether | H, $OCH_3$ |
| (**10**) Artesunate | H, OCO(CH$_2$)$_2$CO$_2$Na |

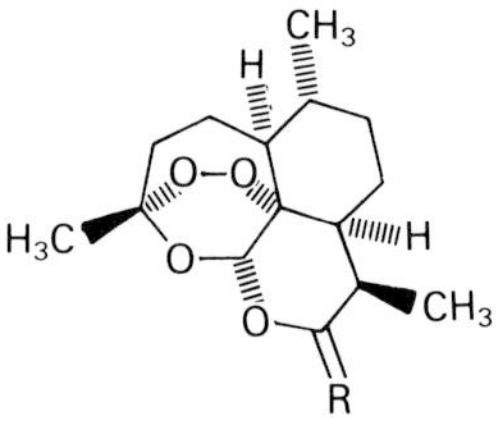

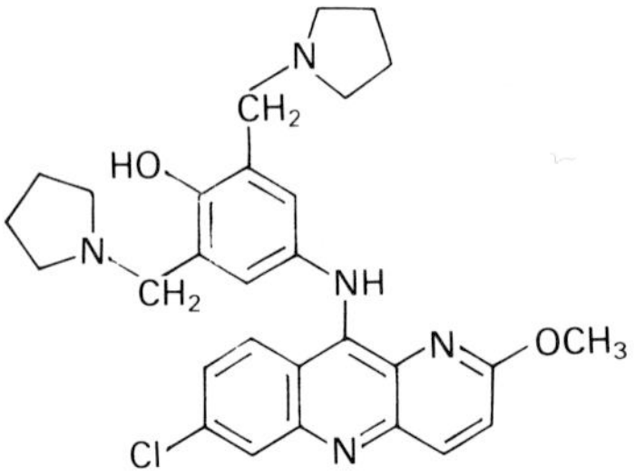

(**11**) Pyronaridine

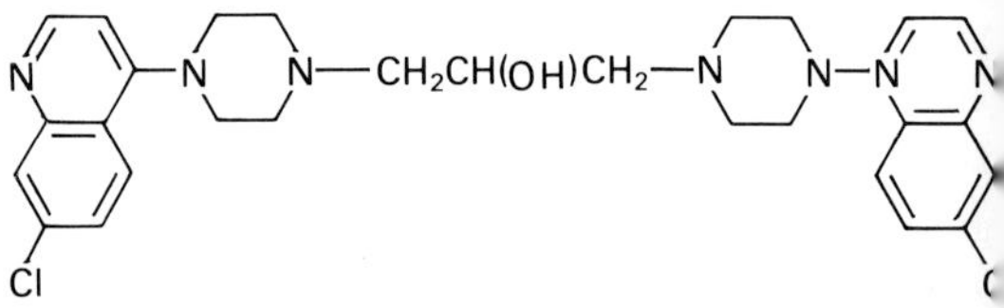

(**12**) Hydroxypiperaquine

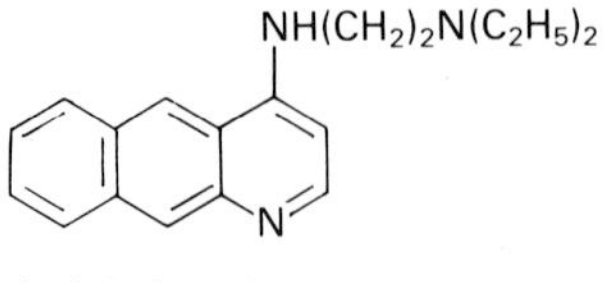

(**13**) Dabequine

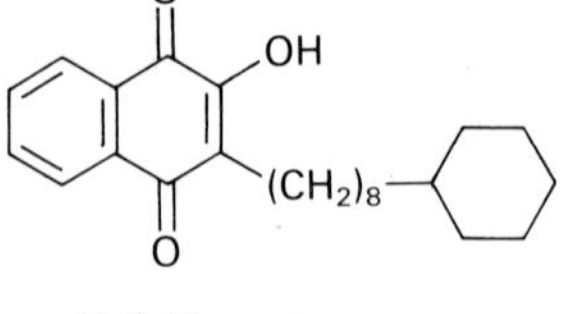

(**14**) Menoctone

shows considerable promise and is undergoing clinical trial; the pyridinemethanol WR 180409 (**7**) is being studied in volunteers.

*(b) Qinghaosu* Traditional remedies derived from plants are still a source of interesting and active substances. Recent studies in China have shown that the active principle, qinghaosu or artemisinine (**8**), of the anipyretic herb *Artemisia annua* is a powerful, rapid antimalarial, active on chloroquine-resistant strains. It is a sesquiterpene peroxide, and derivatives such as the methylether, artemether (**9**), and hemisuccinate, artesunate (**10**), are also active. These compounds are now under clinical trial and are likely to occupy an important place in the chemotherapy of malaria, especially in severe cerebral infections of *P. falciparum* where extreme rapidity of action is essential (Hillier, 1981; Shen and Zuhang, 1983). Synthetic 1,2,4 trioxane derivatives have now been prepared; some of these show high activity against malaria parasites *in vitro*.

*(c) Other new antimalarials* Synthetic antimalarials, also from China, include pyronaridine (**11**) and hydroxypiperaquine (**12**). A new 4-aminoquinoline, dabequine (**13**), has been selected in the USSR and derivatives of menoctone (**14**) have been found with activity against both blood and tissue stages of the parasite. These are all under clinical trial.

Malaria offers many opportunities for drug development. Methods are now available for the culture of human malaria parasites *in vitro*; preliminary screening tests can be carried out rapidly and modes of action investigated. There is a continuing need to keep ahead of the spread of resistance by the introduction of new compounds, not only as schizonticides to control blood infections but also for action against the parasite in the liver cells and against the gametocytes in the blood that carry the infection to the mosquito vector. At present the only compounds available with pronounced activity against tissue parasites and gametocytes are primaquine and related 8-aminoquinolines. These can have serious drawbacks, especially in patients, common in malarious areas, with genetic defects of the erythrocytes such as glucose-6-phosphate dehydrogenase deficiency, in which primaquine causes haemolysis. A new, non-toxic tissue schizonticide is badly needed.

There is no reason why activity against tissue stages and gametocytes should necessarily go hand in hand, and it is likely that substances exist that are lethal to gametocytes although less effective against other stages of the life cycle. Such a drug would be of great value, if effective in a single dose and free from toxic side effects, in campaigns to limit the spread of drug resistance by preventing the infection of the vector mosquito. Methods are now available for the *in vitro* culture of gametocytes and of exoerythrocytic parasites in liver cells; new leads to active compounds should not be long in coming.

### *1.4.2 Trypanosomiasis and leishmaniasis*

*(a) African trypanosomiasis* Human sleeping sickness is a problem in many parts of Africa, appearing as epidemics that often arise as a result of population movements, settlements or incursions into bush areas where parasites of the

*Trypanosoma brucei* complex occur. The organisms are transmitted by tsetse flies and are frequently harboured by wild and domestic animals that provide reservoirs of infection difficult to identify and remove.

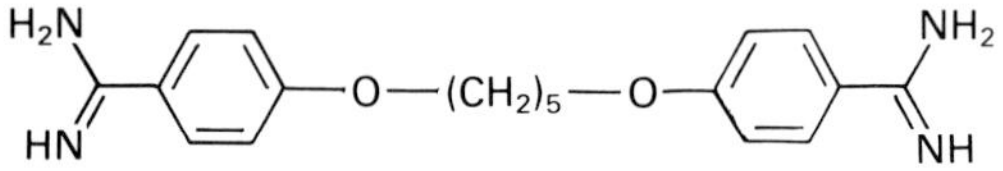

$H_2N$ HN O—$(CH_2)_5$—O $NH_2$ NH

(**15**) Pentamidine

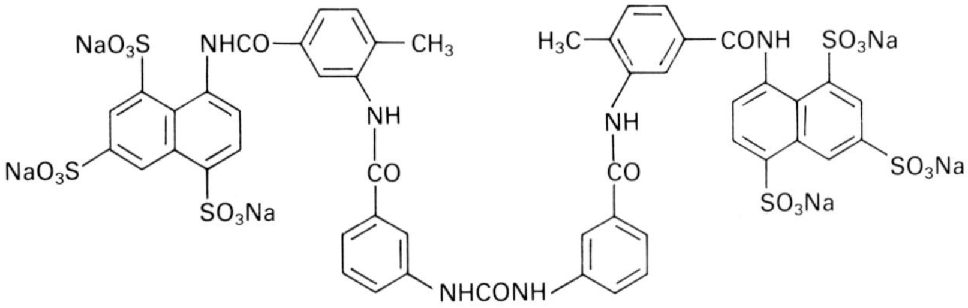

$NaO_3S$ NHCO $CH_3$ $H_3C$ CONH $SO_3Na$ $NaO_3S$ NH CO NH CO $SO_3Na$ $SO_3Na$ $SO_3Na$ NHCONH

(**16**) Suramin

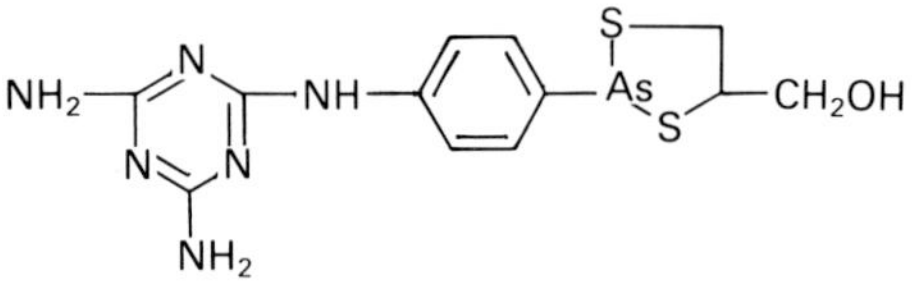

$NH_2$ N N N NH S As S $CH_2OH$ $NH_2$

(**17**) Melarsoprol

$O_2S$ N—N=CH O $NO_2$ $CH_3$

$CH_2CONHCH_2$ N $NO_2$ N

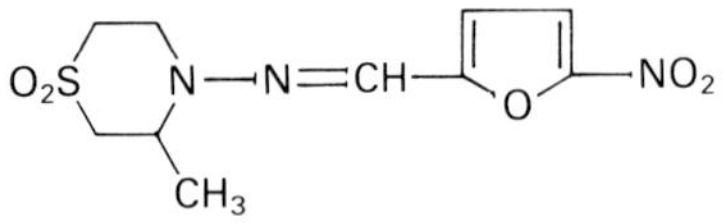

(**18**) Nifurtimox

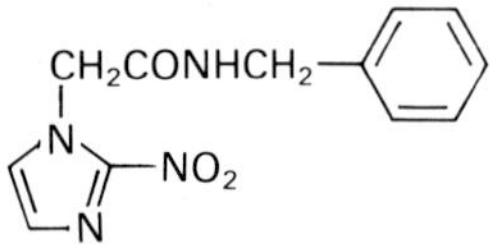

(**19**) Benznidazole

The chemotherapy of human trypanosomiasis is very unsatisfactory and no significant advances have been made for many years. The aromatic diamidine, pentamidine (**15**), can be used as a prophylactic and to treat early infections, and the naphthalenesulphonic acid derivative, suramin (**16**), is of value, but both have unpleasant toxic effects in some patients. If the trypanosomes have entered the central nervous system these drugs are useless because they do not cross the blood–brain barrier. If cerebral involvement has occurred organic arsenic compounds must be used. The most effective available is melarsoprol (**17**), but its use is hazardous because of the danger of arsenical encephalopathy; it is estimated that there is danger to life in 2–5 per cent of patients treated with it. Friedheim, who synthesized melarsoprol forty years ago, has recently prepared a new series of melaminylthioarsenites with lower toxicity and it is to be hoped that these will provide a safer method of treatment. So far, all attempts to produce non-metallic trypanocides with activity against central nervous infections have been unsuccessful.

The pathogenic trypanosomes of domestic animals, *T. congolense*, *T. vivax* and *T. evansi*, are a serious obstacle to agricultural development in Africa, denying vast areas of territory to the raising of livestock.

*(b) South American trypanosomiasis* Chagas' disease, widespread in South America, is caused by *T. cruzi* and is transmitted by reduviid bugs that have become adapted to live in the mud brick walls of human dwellings. The parasite is also found in wild animals that act as reservoir hosts of the disease. The acute stage of infection may be lethal but, more often, it progresses to a chronic stage in which autoimmune factors play a part in the pathology. Damage occurs to the heart muscle cells, in which the parasite divides and multiplies. In some regions the autonomic nerves of the oesophagus or large bowel are affected, causing them to dilate and produce 'megaoesophagus' or 'megacolon'. Two drugs are available for treatment, nifurtimox (**18**) and benznidazole (**19**). These help if given in the acute stage but may have unpleasant side effects and are much less successful in the chronic stage. There is need for better methods of diagnosis at an early stage, for a test of cure which does not depend on feeding bugs on the patient in the hope that they will pick up any trypanosomes that remain and for more effective treatment. Much more needs to be known of the interplay between the parasite and the host's immune response in the pathogenesis of the disease.

*(c) Leishmaniasis* Leishmaniasis, caused by other protozoa of the family Trypanosomidae, takes many forms, from the self-limiting oriental sore of the Middle East to the disfiguring espundia of South America and the lethal kala-azar of Asia and East Africa. These diseases are transmitted by sandflies and the parasites live in the reticuloendothelial cells of the host — the cells that normally engulf and kill invading organisms.

The available treatment with organic pentavalent antimony compounds (sodium stibogluconate (**20**) and meglumine antimonate) leaves much to be desired. A lengthy series of injections is needed, most of the dose being rapidly excreted in the urine, and cure is not always attained, especially in the South American mucocutaneous forms of the disease.

If the antimonial is encapsulated in liposomes of appropriate size and formulation, its effect on experimental leishmaniasis is much greater than if given in aqueous solution; the particles are taken up by the phagocytes that contain the parasites and the release of the drug kills the target organisms. However, liposomes are not easy to prepare to an acceptable standard and at present they have a short shelf-life. For this and other reasons they have not yet become available for use in the field.

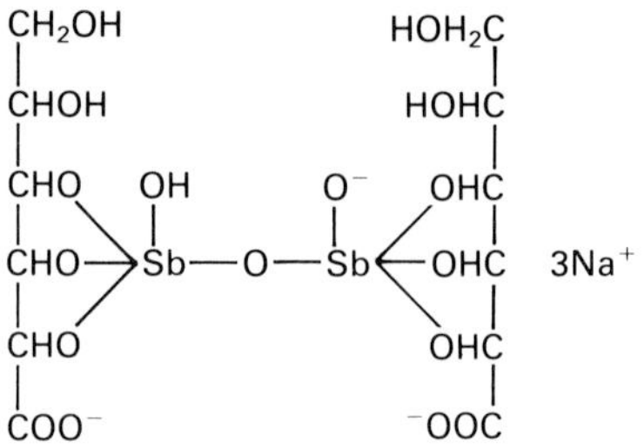

(**20**) Sodium stibogluconate

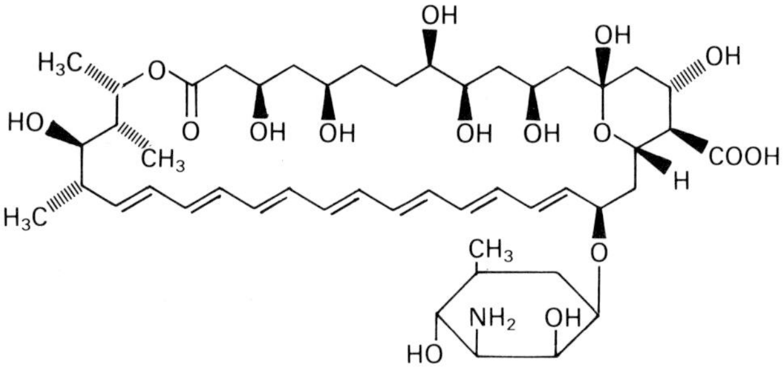

(**21**) Amphotericin B

Amphotericin B (**21**) also has activity in leishmaniasis but has unpleasant side effects and the cost is high. There is need for much better and more effective drugs for the treatment of all forms of leishmaniasis.

### *1.4.3 Schistosomiasis*

Schistosomes are small trematode worms that live in pairs in the veins of the bladder (*Schistosoma haematobium*), the large bowel (*S. mansoni*) or more widely distributed through the veins of the abdominal organs (*S. japonicum*). For many years the only method of treatment was with trivalent antimony compounds given by repeated intravenous injection. A breakthrough occurred with the introduction of the thioxanthone compound, lucanthone, and its active metabolite, hycanthone (**22**), followed by the organophosphorus anticholinesterase, metriphonate (**23**), and the nitrothiazole, niridazole (**24**). A further important advance came when the veterinary anthelmintics, oxamniquine (**25**) and praziquantel (**26**), were found to be effective in human schistosomiasis. Praziquantel is active against all species of schistosome pathogenic to man and its side effects are mild, transient and non-specific (Andrews *et al.*, 1983). A single dose frees the patient from most of his worm load and this makes it possible, for the first time, to organize mass treatments with some hope of controlling the disease. Oxamniquine is active against *S. mansoni*, but much less so on other

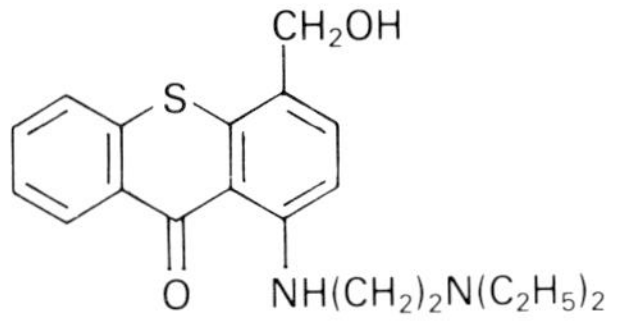

(**22**) Hycanthone

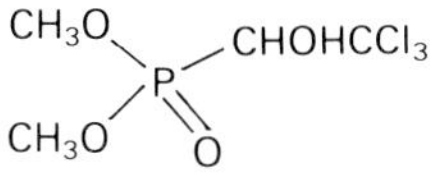

(**23**) Metrifonate

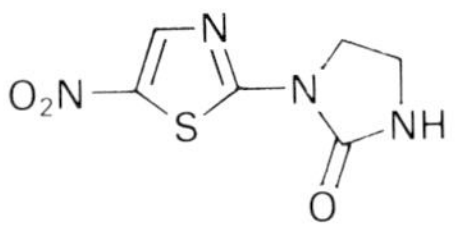

(**24**) Niridazole

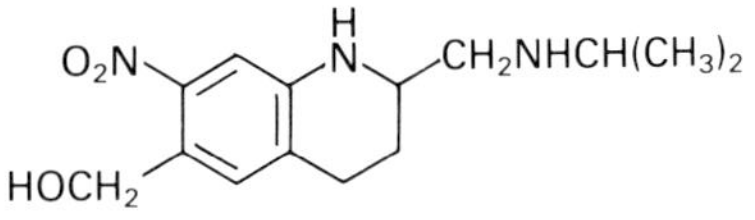

(**25**) Oxamniquine

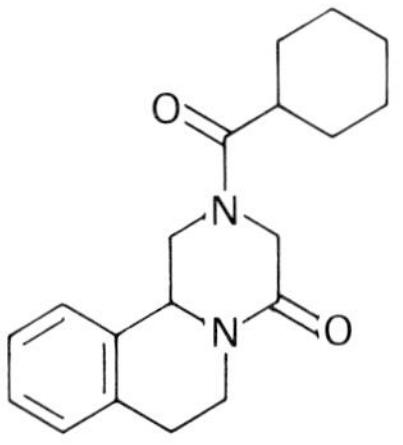

(**26**) Praziquantel

species; there are differences in susceptibility among various geographic strains of worm and the drug sometimes produces dramatic side effects.

Efforts over many years to control schistosomiasis by killing the snail intermediate hosts of the parasite with molluscicides have not met with lasting success. Molluscicides affect fish and other aquatic creatures in the environment and snails can rapidly repopulate water even if only a single one is left deep in the mud. However, if used sensibly, the combined treatment of patients and the control of snail populations could be successful in breaking transmission.

### *1.4.4 Filariasis*

There are eight species of filarial nematode that infect man; the most important are *Onchocerca volvulus* that causes river blindness in Africa and parts of Central and South America, and *Wuchereria bancrofti* and species of *Brugia* that cause lymphatic filariasis, leading to elephantiasis in Africa and Asia.

*(a) Onchocerciasis* A vast project, the Onchocerciasis Control Programme (OCP) is now in operation to kill the blackfly (*Simulium*) that transmits the disease, by treating breeding sites in the rivers of the Volta River basin in West Africa with larvicide. However, there are signs that the fly is becoming resistant to the synthetic chemicals used, and trials are being made of biological control

methods using *Bacillus thuringiensis*. Also, many thousands of human cases remain and the *Onchocerca* worms that inhabit nodules in the skin and elsewhere in the body are long lived, producing larvae (microfilariae) for many years. An urgent search has now begun for effective drugs to kill the adult worms.

Until recently, the only drugs available were suramin (**16**), which may cause unexpected, sometimes fatal, side effects, and diethylcarbamazine (**27**), which

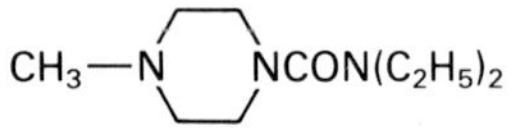

(**27**) Diethylcarbamazine

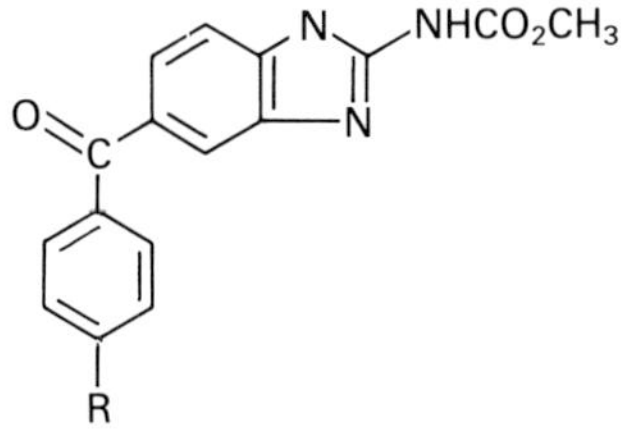

(**28**) Mebendazole, R = H

(**29**) Flubendazole, R = F

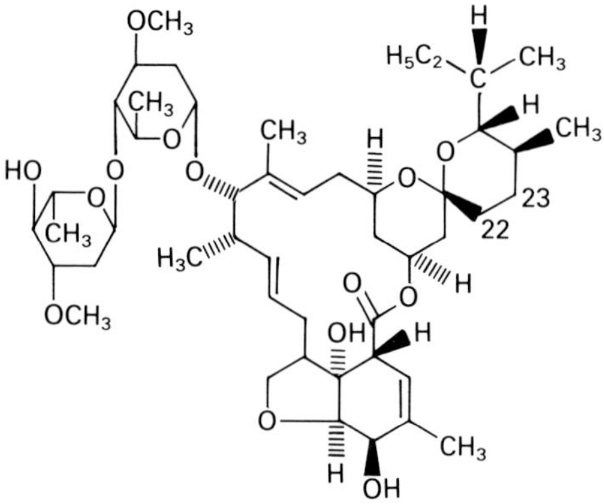

22,23-Dihydroavermectin $B_{1a}$ (80%)

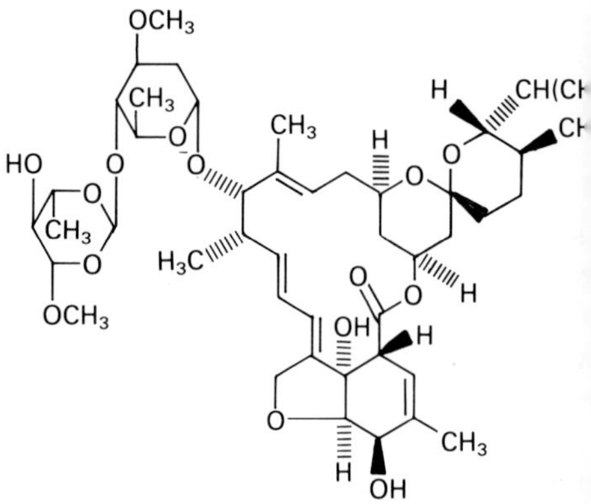

22,23-Dihydroavermectin $B_{1b}$ (20%)

(**30**) Ivermectin

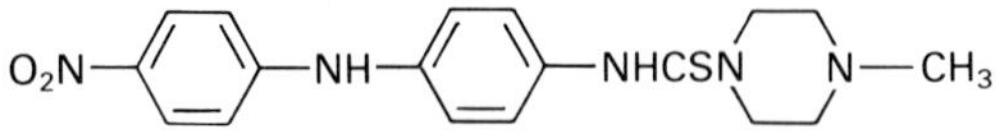

(**31**) CGP 6140

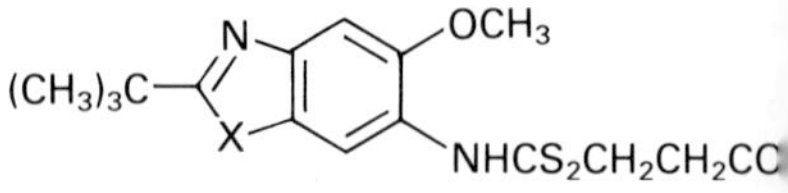

(**32**) CGP 20376, X = S

(**33**) CGP 24914, X = O

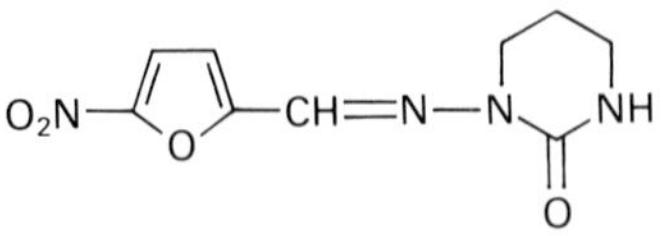

(**34**) Furapyrimidone

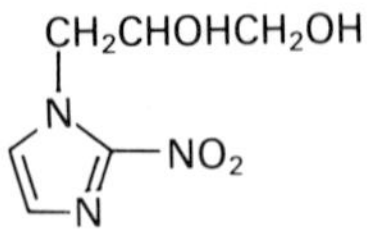

(**35**) Desmethylmisnidazole

by its action on the microfilariae that swim in the tissue spaces often causes serious allergic (Mazzotti) reactions. Microfilariae are often found in the eyes and eventually cause blindness; if diethylcarbamazine is given the allergic response may make the condition worse, not better. As a result of research stimulated by the WHO Special Programme, several promising new drugs are now on trial. Benzimidazole derivatives, introduced as veterinary anthelmintics, have an action upon the reproductive system of female *Onchocerca*, causing the eggs in the uterus to stop development at the blastula stage. Mebendazole (**28**) has to be given in large, frequent doses because little is absorbed; its effect lasts for several months and then some of the worms recover. A much better, longer-lasting action occurs if flubendazole (**29**) is given by intramuscular injection, but the injection is painful and attempts are now being made to devise a more acceptable formulation.

Perhaps the most promising new development is the discovery that ivermectin (**30**) has high activity against *Onchocerca* microfilariae. Ivermectin is a mixture of macrocyclic lactones derived from *Streptomyces avermitilis* and was introduced as veterinary anthelmintic; it also acts against arthropod ectoparasites. It causes the disappearance of microfilariae from the skin after a single dose given to patients with onchocerciasis and there is little or no Mazzotti reaction. All other drugs that affect microfilariae cause a troublesome allergic response and the reason why ivermectin does not is an intriguing problem for future study.

Other promising new drugs are three macrofilaricidal compounds synthesized by Ciba-Geigy (**31**, **32**, **33**) and Friedheim's new melaminylthioarsenites. If the arsenical compounds are shown to be free from toxicity in patients with potentially fatal trypanosomiasis, it may be justifiable to test them in human filarial infections. Further compounds with promising activity in experimental animal screens are furapyramidone (**34**) and desmethylmisnidazole (**35**) (Goodwin, Ottesen and Southgate, 1984).

*(b) Lymphatic filariasis Wuchereria bancrofti* and species of *Brugia* are long filarial nematodes that live in the lymphatics. They are transmitted by mosquitos that pick up microfilariae from the blood of infected patients. Sometimes the worms cause little apparent pathology but in some patients the lymphatics become occluded by local tissue reaction around the worms and a leg, arm or sometimes the breast or scrotum swells to a great size and becomes permanently indurated. Diethylcarbamazine is effective in the early stages and it is possible to adjust the dose so that allergic disturbances due to the death of microfilariae are minimal. If treatment is given adequately the adult worms die. If sufficient control can be exercised, communities may be cured and protected by distribution of diethylcarbamazine in cooking salt. It is to be hoped that newer drugs will prove to be more rapid in action and give rise to less side effects than diethylcarbamazine.

Filariasis still offers a challenge because very large numbers of people are at risk; little is understood of the immune response that leads some people and not others to be seriously affected by the infection. There is an immediate need, especially with onchocerciasis in the OCP to diagnose infections at an early stage and to ascertain when the adult worms have been killed by chemotherapy (WHO 1984b).

## 1.5 Future prospects for drug development

The existence of many thousands of compounds prepared for various purposes in the laboratories of the pharmaceutical industry offers an opportunity for the discovery of new leads towards improved treatment of tropical disease. The WHO Special Programme has sought and obtained valuable cooperation from companies that have either carried out screening themselves or have made available, under code numbers if they so wished, substances for testing in units set up and financed by the Programme in university departments, polytechnics and other laboratories. A system of mutual trust and confidence has been built up since the programme began in 1977 and useful advances have been made. In several laboratories, in universities, polytechnics, and in industry, lead-directed programmes of synthesis have been supported by the WHO, and financial assistance has been made available to carry out toxicological tests on promising compounds. Arrangements have also been made for the conduct of carefully controlled clinical trials in suitable hospitals in endemic areas; these are monitored by the WHO. For the discovery, development and production of new drugs the pharmaceutical industry is essential; the WHO can greatly assist in the process and in the trial and deployment of promising compounds for diagnosis and treatment. It is to be hoped that this useful collaboration will be continued and extended.

### *1.5.1 Basic studies*

The more or less random screening of available compounds sooner or later runs out of steam; although the method has yielded useful antiparasitic drugs, some of them prepared with entirely different purposes in mind, there is a need to know as much as possible about the biochemical processes of the parasite in comparison with those of the host, to facilitate the rational design of new compounds which would selectively disrupt these processes. Knowledge of parasite biochemistry is accumulating (Gutteridge and Coombes, 1977; Barrett, 1981) but far too little is understood of the ways in which the available active substances work. There are a few exceptions — pyrimethamine inhibits dihydrofolate reductase and is potentiated by sulphonamides that inhibit pteroate synthetase in the same metabolic pathway. The organophosphorus anthelmintic, haloxon (**36**), works because it binds firmly with the cholinesterase of parasitic nematodes but only loosely and briefly with the corresponding enzyme of the mammalian host.

Stratagems of this kind do not always succeed; salicylhydroxamic acid (**37**) is an inhibitor of L-*d*-glycerophosphate oxidase, an unusual key enzyme in the glycolysis of African trypanosomes, and it is potentiated by glycerol. Its activity *in vivo*, with or without glycerol, is disappointing. However, the discovery of activity

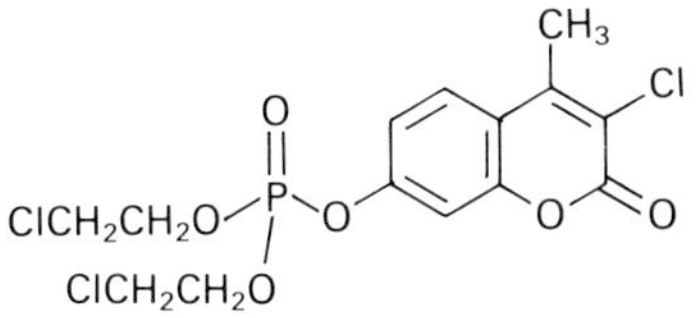

(**36**) Haloxon, BAN, NFN

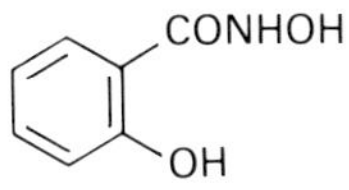

(**37**)

in any new chemical type opens up exciting possibilities and leads to new information that can help in the unravelling of host/parasite biochemistry.

Ivermectin, with its astonishingly wide range of activity against parasitic helminths and arthropods, is believed to interfere with the release of γ-aminobutyric acid (GABA), a vital neurotransmitter in these creatures.

Basic studies of amino acid and polyamine metabolism of trypanosomes are providing new ideas for drug development. Trypanosomes themselves are now a popular subject of study by molecular biologists who have become fascinated by the ability of these organisms to evade the immune responses of the mammalian host by varying the structure of the glycoprotein coats they secrete around themselves when they leave the tsetse vector. The field is open for a new kind of chemotherapy to interfere with this process of antigenic variation, or with the ability of the trypanosome to produce the glycoprotein coat or fix it to its surface membrane. The field is also open for a study of the reproductive processes of parasitic worms. The action of benzimidazoles is believed to be on the tubulin structure of the parasite; their specific action on the embryos of filarial worms has led to a search for information on nematode reproduction. Very little is known — it is not even clear whether steroids are involved in these processes, although this would seem likely since Nature has found the steroid nucleus to provide a useful template in the multiplication of even the most primitive unicellular organisms. If filarial reproduction does involve steroids, the way is open for the synthesis of antagonists or modifiers with a structure unlikely to affect the host. A nematode contraceptive that would prevent a female *Onchocerca* from producing her regular 6000 larvae a day would be of considerable value as a medicine.

In malaria, the antifolate drugs interfere with thymidylate synthesis, but there is an opportunity to design other antimetabolites. For example, in *Plasmodium falciparum*, orotate is metabolized first to orotidine-5′-monophosphate and then to uridylate monophosphate. This differs from the mammalian route of pyrimidine synthesis and is a possible target for chemotherapy.

Malaria parasites, at a critical stage in their life cycle, are released from one red cell and have to enter another. Several studies have shown that the erythrocyte sialoproteins glycophorin A and B are receptors for the malarial merozoites. Highly selective binding to these carbohydrate determinants occurs and a sugar-binding protein has been identified on the parasite surface. The process could provide a sensitive target for chemotherapy of an entirely new type.

These are only a few of the developing opportunities for chemotherapeutic research.

### *1.5.2 New types of drug formulation*

Drug formulation techniques have been changing fast. Long-acting products can now be made by preparing salts of low solubility and incorporating them in a suitable carrier; active compounds may be fixed in an inactive polymer so that the action of the drug develops only after biotransformation of the carrier. Drugs can also be incorporated into biologically inert matrices that delay or extend their release at an accurately predetermined rate.

Liposomes have been shown in experimental infections to carry drugs to appropriate targets. Increased activity is shown by antimonials in leishmaniasis and, in malaria, primaquine has a lower toxicity for the same activity if it is incorporated into liposomes. Liposomes are also effective as a drug delivery system in schistosomiasis because they accumulate in the liver and in the reticuloendothelial cells. The development of monoclonal antibodies with a high specificity for various parasite life stages may well provide a novel means of carrying drugs to their targets.

### *1.5.3 Opportunities*

The development of new medicines for parasitic diseases is not easy but is badly needed and offers an exciting challenge to chemists and biomedical scientists in all disciplines. More knowledge of the physiology and biochemistry of the parasites is needed, together with an understanding of the mechanisms of pathogenicity to their hosts. These are promising areas for research and new techniques are now available for rapid advances to be made. New drugs are coming into use, but there will be continuing problems because of the different forms that the diseases take in different countries and different peoples, and because of the menace of drug resistance.

## 1.6 References

1. P. Andrews, H. Thomas, R. Polk, and J. Seubert, (1983). *Med. Res. Rev.*, **3**, 147–200.
2. J. Barrett, (1981). *Biochemistry of Parasitic Helminths*, Macmillan, London.
3. L.G. Goodwin, E.A. Ottesen, and B.A. Southgate, (1984). *Trans. R. Soc. Trop. Med. Hyg.*, **78**, Supplement.
4. W.E. Gutteridge, and G.H. Coombes, (1977). *The Biochemistry of Parasitic Protozoa*, Macmillan, London.
5. K. Hillier, (1981). In *Drugs of the Future*, VI, Vol. 37, Prous, Barcelona.
6. W. Peters, and W.H.G. Richards, (Eds.) (1984). *Handbook of Experimental Pharmacology*, Vol. 68/II, *Antimalarial Drugs*, Springer, Berlin.
7. C-C. Shen, and L-G. Zuhang, (1983). *Med. Res. Rev.*, **4**, 47–86.
8. T.R. Sweeney, (1981). *Med. Res. Rev.*, **1**, 281–301.
9. UNDP/World Bank/WHO (1984). Special Programme for Research and Training in Tropical Diseases; Sixth Programme Report, 1 July 1981–31 December 1982.
10. UNDP/World Bank/WHO (1983). Modern Design of Antimalarial Drugs (Eds. W. Wernsdorfer and P. Trigg).
11. WHO (1984a). Chemotherapy of Malaria, Report of a WHO Scientific Group.
12. WHO (1984b). Fourth Report of the Expert Committee on Filariasis, Technical Report Series, WHO.
13. (WHO documents are obtainable by application to: The World Health Organization, 1211 Geneva 27, Switzerland.)

# 2 Chemical targets in the cell envelopes of the leprosy bacillus and related mycobacteria

David E. Minnikin
Department of Organic Chemistry, The University, Newcastle upon Tyne NE1 7RU, UK



## 2.1 Introduction

The conquest of mycobacterial diseases continues to present a formidable challenge. In particular, leprosy remains an apparently insurmountable problem in parts of Africa, Central America and the Indian subcontinent. Drug regimens are available for the successful treatment of leprosy but their administration often necessitates a very high standard of clinical and nursing expertise which is not always available where required. The mode of action of the principal antileprosy agent, diaminodiphenylsulphone (dapsone), is well understood (Winder, 1982; Kulkarni and Seydel, 1983), the action being on folate-synthesizing enzymes. The sensitivity of *Mycobacterium leprae* and an environmental organism, *M. lufu*, is attributed to the high affinity of the drug for the dihydropteric acid synthetase (Kulkarni and Seydel, 1983).

Rifampin (rifampicin) is also included in antileprosy drug regimes (Hooper and Purohit, 1983; Godal and Levy, 1984) and it inhibits protein synthesis at the transcription level (Winder, 1982; Hawkins, 1984). Clofazimine and prothionamide are also found in antileprosy regimes (Hooper and Purohit, 1983; Godal and Levy, 1984). The former was thought to interfere with respiration (Winder, 1982) but recent studies in mice have shown increases in the activity of various lysosomal enzymes and the amount of labelled immune complexes phagocytosed by macrophages (Sarracent and Finlay, 1984). The precise mode of action of thioamides, such as prothionamide, is unknown; reports suggest some interference with mycolic acid and other lipid synthesis (Hooper and Purohit, 1983). Chaulmoogra oil, traditionally used in leprosy therapy, contains ω-cyclopentenyl fatty acids, and these have some activity against *M. leprae* (Levy, 1975) and other mycobacteria (Winder, 1982). It has been demonstrated that chaulmoogric (ω-cyclopentenyltridecanoic) acid is incorporated into the cellular phospholipids and triacylglycerols of *M. vaccae*, suggesting an interference with membrane function (Goucher and Cabot, 1981). A brief summary of the current status and future strategy in the search for the antileprotic drugs has been given by Hooper (1985).

Other mycobacterial diseases with tropical significance include tuberculosis (Bates, 1984) and infections caused by *M. ulcerans* (Pattyn, 1984). Rifampin and isoniazid are the major drugs in the current multidrug regimens used for the treatment of tuberculosis, and these regimens may also include ethambutol and pyrazinamide. Streptomycin and *p*-aminosalicylic acid are still widely used, particularly in developing countries, and cycloserine and capreomycin are also effective (Raleigh, 1984). Chemotherapy of *M. ulcerans* has not been thoroughly studied but clofazimine, rifampin and dapsone are inhibitory, the latter being the least effective (Pattyn, 1984).

The drugs mentioned above may be classified into different groups depending on their mode of action. As summarized by Winder (1982) and Hawkins (1984), certain drugs such as capreomycin, rifampin and streptomycin interfere with protein biosynthesis and their effectiveness is not restricted to mycobacteria. Other drugs, however, inhibit the biosynthesis of components of the cell envelope. Cycloserine, for example, interferes with the action of enzymes involved in peptidoglycan biosynthesis and ethambutol and isoniazid have an effect on lipid synthesis. The latter two drugs, however, also have more general effects on nucleic acid and other cellular metabolism.

The physiology of the cell envelopes of mycobacteria is very unusual in having an outer membrane, incorporating complex free lipids, associated with an arabinogalactan–mycolic acid matrix (Minnikin, 1982). As will be outlined below, many of the lipid structures are species specific and their biosynthetic pathways present possible targets for antibiotics. The purpose of the present review is to attempt to categorize the wide variety of envelope chemical targets so that the potential of existing and new antimycobacterial drugs can be viewed in a broader context and their application developed in a more systematic manner.

## 2.2 Mycobacterial cell envelopes

Early studies on the lipid composition of *M. tuberculosis* and other mycobacteria revealed the presence of many unusual lipids not found in any other living organisms (Anderson, 1941). The essential nature of these novel compounds slowly emerged as chronicled in the monography by Asselineau (1966). The distribution of mycobacterial lipid types has been surveyed recently (Asselineau and Asselineau, 1978a, 1978b; Goren and Brennan, 1979; Minnikin and Goodfellow, 1980; Minnikin, 1982; Dobson *et al.*, 1985) and the composition of cell wall polymers detailed by Lederer *et al.* (1975) and Draper (1982). The chemical anatomy of mycobacterial envelopes will be outlined in the following sections and an updated version of a chemical model of the mycobacterial outer membrane (Minnikin, 1982) will be given. The main structural unit in the envelope of mycobacteria is a conventional peptidoglycan with an attached arabinogalactan polysaccharide. High molecular weight mycolic acids are covalently linked to the arabinogalactan and complex outer membrane free lipids are associated with this mycolic acid matrix. The innermost organelle in the mycobacterial envelope is the essential plasma membrane.

### *2.2.1 Peptidoglycan–arabinogalactan*

Mycobacterial peptidoglycan corresponds to one of the commonest types found in bacteria, being composed of a polysaccharide formed from *N*-acetylglucosamine and muramic acid (3-*O*-lactylglucosamine) crosslinked with an L-alanyl-D-isoglutamyl-*meso*-diaminopimelyl-D-alanyl tetrapeptide (Figure 2.1) (Draper,

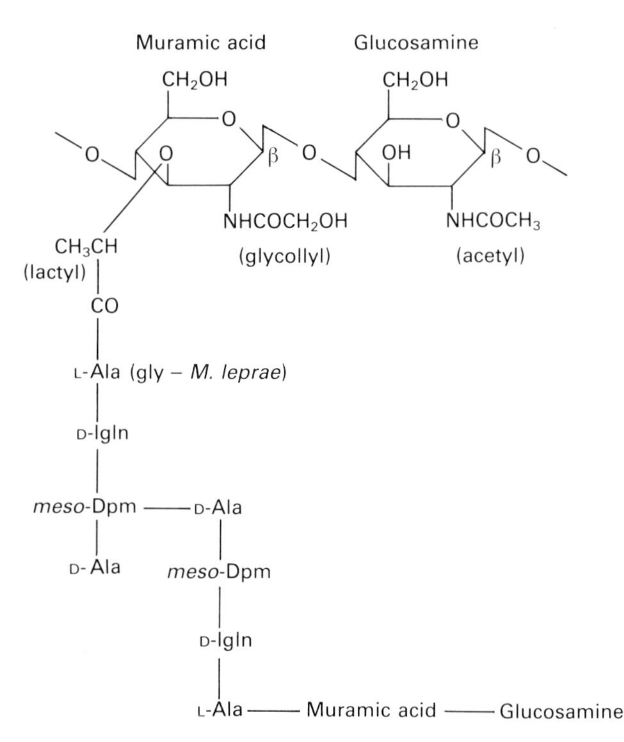


**Figure 2.1** Essential structure of mycobacterial peptidoglycan. Abbreviations: Ala, alanine; Igln, isoglutamine; Dpm, diaminopimelic acid.

1982). The muramic acid is, however, *N*-glycollated rather than having the more usual *N*-acetyl substituent (Figure 2.1) and a proportion of the crosslinks may be between two residues of diaminopimelic acid instead of between diaminopimelic acid and D-alanine (Draper, 1982). Muramic acids acylated with glycollic acid are also encountered in the walls of other actinomycetes (Draper, 1982; Minnikin and O'Donnell, 1984). The peptidoglycan of *M. leprae* is, however, distinguished from those of other mycobacteria by the replacement of L-alanine (Figure 2.1) by glycine (Draper, 1982).

Arabinogalactan polysaccharides in the mycobacterial wall are considered to be linked by phosphodiester bonds to a proportion of the 6-positions of the muramic acid residues (Lederer *et al.*, 1975). The most complete model for the structure of the arabinogalactan has been developed by Misaki, Seto and Azuma (1974) (Figure 2.2), but uncertainties remain regarding whether the galactoses are 1→4-linked galactopyranoses (Misaki, Seto and Azuma, 1974) or 1→5-linked galactofuranose units (Vilkas *et al.*, 1973). Mycolic acid residues are attached to terminal arabinofuranoses (Figure 2.2).

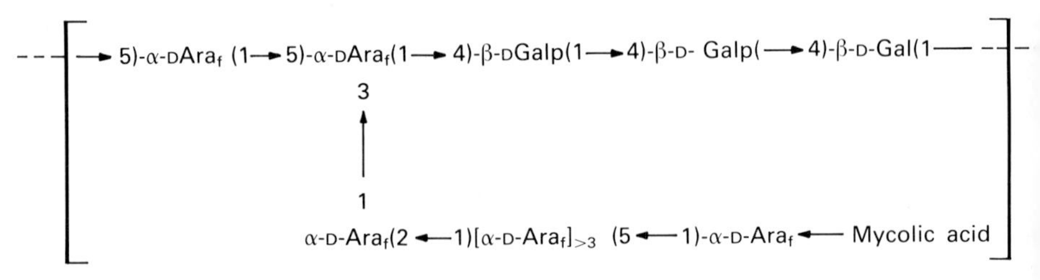


**Figure 2.2** Possible structure of mycolylarabinogalactan. Abbreviations: D-Ara$_f$, D-arabinofuranose; D-Galp, D-galactopyranose.

The antitubercular drug ethambutol has been considered to interfere with lipid synthesis (Winder, 1982; Hawkins, 1984), an effect on the balanced incorporation of mycolic acids into the cell envelope being noted (Takayama *et al.*, 1979; Kilburn and Takayama, 1981). In a recent study (Takayama and Kilburn, 1984) it has been proposed that the primary mode of action of ethambutol is the inhibition of arabinogalactan synthesis. In particular the precise site of action appeared to be on the pathway between the conversion of D-glucose to D-arabinose and the transfer of the latter into the arabinogalactan. As noted in the introduction, cycloserine is known to inhibit peptidoglycan biosynthesis enzymes (Winder, 1982; Hawkins, 1984). A recent report (Cynamon and Palmer, 1983) has shown that amoxicillin combined with clavulanic acid has significant activity against *M. tuberculosis*, suggesting that lactamases may be at least partially responsible for the resistance of mycobacteria to penicillin antibiotics.

### *2.2.2 Mycolic acids*

Mycolic acids are high molecular weight (60–90 carbon) 3-hydroxy fatty acids with

**Table 2.1** Structures of representative mycolic acids

| | |
|---|---|
| α | $CH_3(CH_2)_l\,X(CH_2)_mY(CH_2)_n\overset{HO}{C}H\overset{COOH}{C}H(CH_2)_xCH_3$ |
| | X = *cis* $—CH=CH—$, $—\overset{CH_2}{CH—CH}—$; Y = X or *trans* $—CH=CH\overset{CH_3}{C}H—$ with $(CH_2)_{m-1}$ |
| | $l = 15, 17, 19;\ m = 14, 16;\ n = 11, 13, 15, 17;\ x = 19, 21, 23$ |
| α′ | $CH_3(CH_2)_lCH=CH(CH_2)_m\overset{HO}{C}H\overset{COOH}{C}H(CH_2)_xCH_3$ (*cis*) |
| | $l = m = 17;\ x = 21$ |
| Methoxy | $CH_3(CH_2)_l\overset{H_3C}{C}H\overset{OCH_3}{C}H(C_yH_{2y-2})\overset{HO}{C}H\overset{COOH}{C}H(CH_2)_xCH_3$ |
| Keto | $CH_3(CH_2)_l\overset{CH_3}{C}H\overset{O}{\overset{\|}{C}}(C_yH_{2y-2})\overset{HO}{C}H\overset{COOH}{C}H(CH_2)_xCH_3$ |
| Epoxy | $CH_3(CH_2)_l\overset{H_3C}{C}H\overset{O}{CH—CH}(C_yH_{2y-2})\overset{HO}{C}H\overset{COOH}{C}H(CH_2)_x$ (*trans*) |
| Wax ester | $CH_3(CH_2)_l\overset{CH_3}{C}HO\overset{O}{\overset{\|}{H}}(C_yH_{2y-2})\overset{HO}{C}H\overset{COOH}{C}H(CH_2)_x$ |
| | $l = 15, 17;\ y = 32–39;\ x = 19, 21, 23$ |

a large n-alkyl side chain at the 2-position. Many different structural types of mycolic acids have been characterized from mycobacteria and related actinomycete taxa such as corynebacteria, norcardiae and rhodococci (Minnikin and O'Donnell, 1984). The mycobacterial mycolic acids are the most complex in structure, having a variety of different oxygen functions and skeletal features. A selection of mycobacterial mycolic acids is given in Table 2.1. In contrast, non-mycobacterial mycolic acids have relatively simple structures varying only in their overall size (20–70 carbons), number of cis double bonds and length of the side chain (Minnikin and O'Donnell, 1984). A number of characteristic combinations of mycolic acid types are encountered in mycobacteria and an appropriate selection is given in Table 2.2; more detailed information is available elsewhere (Minnikin and Goodfellow, 1980; Minnikin, 1982; Daffé *et al.*, 1983; Minnikin *et al.*, 1984, 1985c; Dobson *et al.*, 1985).

A common feature in the structures of mycobacterial mycolic acids is the

**Table 2.2** Distribution of mycobacterial mycolic acids[a]

| Pattern of structural types[b] | Mycobacterial species |
|---|---|
| α only | *M. fallax*, *M. triviale* |
| α, α′ | '*M. borstelense*', *M. chelonae* |
| α, α′, methody | *M. agri* |
| α, α′, epoxy | *M. chitae*, *M. farcinogenes*, *M. fortuitum* '*M. giae*', '*M. peregrinum*', *M. senegalense*, *M. smegmatis* |
| α, α′, keto | *M. simiae* |
| α, α′, keto, wax ester | *M. chubuense*, *M. duvalii*, *M. gilvum*, *M. parafortuitum*, *M. vaccae* |
| α, keto, wax ester | *M. aichiense*, *M. aurum*, *M. avium*, *M. flavescens*, *M. gadium*, *M. gallinarum*, *M. intracellulare*, *M. lepraemurium*, *M. nonchromogenicum*, *M. neoaurum*, '*M. novum*', *M. paratuberculosis*, *M. phlei*, *M. rhodesiae*, *M. scrofulaceum*, *M. terrae*, *M. tokaiense*, *M. xenopi* |
| α, keto | *M. bovis* BCG, *M. leprae* |
| α, keto, methoxy | *M. asiaticum*, *M. bovis*, *M. gastri*, *M. gordonae*, *M. kansasii*, *M. marinum*, *M. microti*, *M. tuberculosis*, *M. szulgai*, *M. ulcerans* |
| α, α′, methoxy, keto | *M. thermoresistibile* |
| α, keto, methoxy, wax ester | *M. komossense* |

[a]Simplified from Dobson *et al.* (1985) and Minnikin *et al.* (1984, 1985c).
[b]See Table 2.1 for examples of mycolic acid structures.

regular spacing, in the main chain, of the oxygen functions, double bonds and cyclopropane rings (Table 2.1). Considerations of possible arrangements for the cooperative interactions of such fatty acids lead to the proposal that mycolic acid chains form a monolayer membrane covalently bound to the basal cell wall arabinogalactan (Minnikin, 1982). Such a mycolic acid matrix would provide an anchorage for a range of complex free lipids leading to a functional bilayer outer membrane in mycobacteria. Details of the suggested arrangement of the lipid components in the mycobacterial outer membrane will be included in a later section (2.2.5) on the overall chemical model of the mycobacterial envelope. It is clear, however, that mycolic acids are essential structural building blocks in the envelopes of mycobacteria and it is important, therefore, to consider their biosynthesis.

The biosynthesis of mycolic acids has been the subject of much discussion but only recently has detailed evidence become available. Early studies demonstrated that, in *Corynebacterium diphtheriae*, 32-carbon mycolic acids were formed by condensation and reduction of 16-carbon fatty acids (Gastambide-Odier and Lederer, 1960) and similar condensations were established for the mycolic acids from *Nocardia asteroides* and *Mycobacterium smegmatis* (Etémadi, 1967). The only detailed extended investigation of mycobacterial mycolic acid biosynthesis has been performed by Takayama and coworkers for *Mycobacterium tuberculosis*

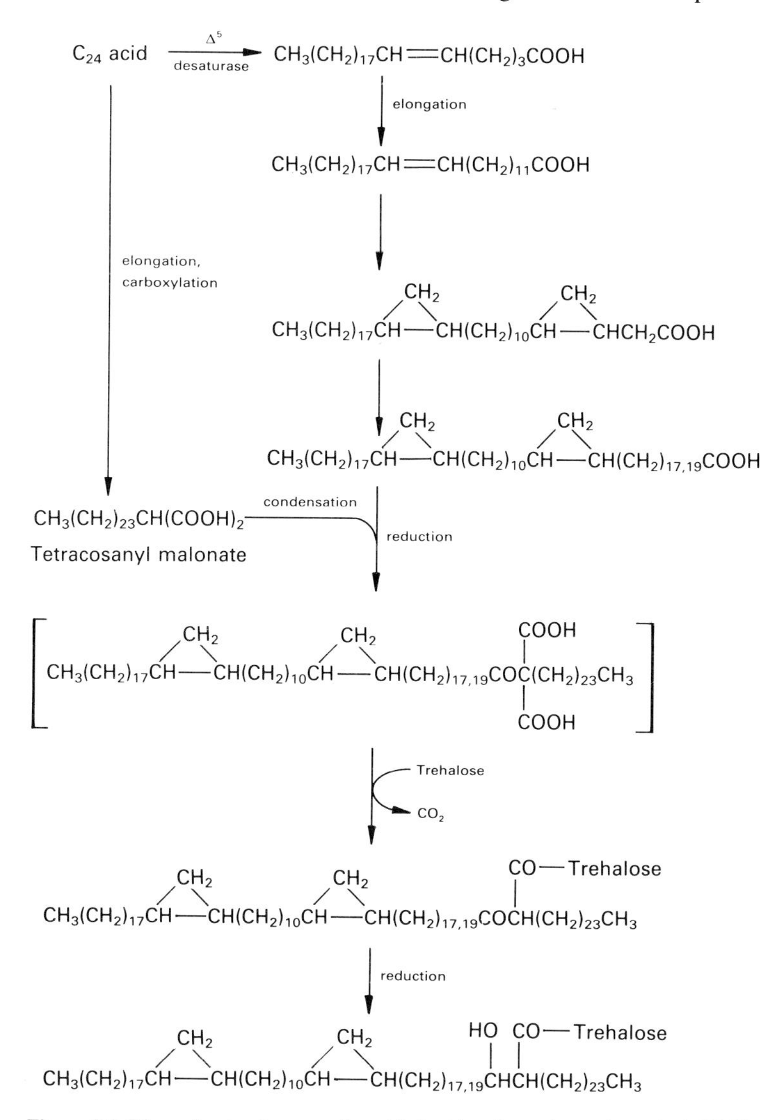


**Figure 2.3** Biosynthesis of α-mycolic acids in *Mycobacterium tuberculosis* H37Ra.

H37Ra, as reviewed by Minnikin (1982) and Takayama and Qureshi (1984). The general pathway for the biosynthesis of the α-mycolic acids in *M. tuberculosis* H37Ra is outlined in Figure 2.3. The 24-carbon acid-Δ-5 desaturase reaction was considered to be one-step sensitive to the antitubercular drug isoniazid (Winder, 1982; Takayama and Qureshi, 1984) but a detailed mechanism for the inhibition and the resistance of certain mycobacteria has not been proposed. Isoniazid also appears to inhibit the elongation of fatty acids having more than 30 carbons (Figure 2.3) (Takayama and Qureshi, 1984). In a recent study (Qureshi,

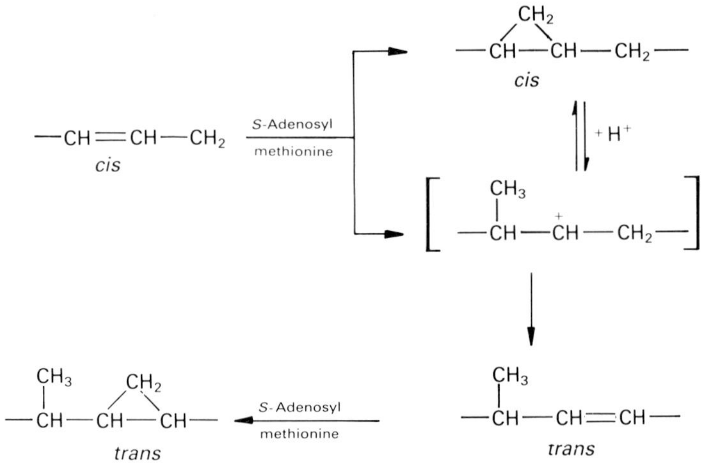

—CH═CH—CH₂ (cis) —[S-Adenosyl methionine]→ —CH(CH₂)—CH—CH₂— (cis cyclopropane) ⇌ (+H⁺) [—CH(CH₃)—C⁺H—CH₂—] → —CH(CH₃)—CH═CH— (trans) —[S-Adenosyl methionine]→ —CH(CH₃)—CH(CH₂)—CH— (trans cyclopropane)

**Figure 2.4** Transformation of olefin linkages in mycolic acids.

Sathyamoorthy and Takayama, 1984) a cell-free extract of *M. tuberculosis* was capable of synthesizing meromycolate-like $C_{48}$ to $C_{56}$ fatty acids, indicating that some of the enzymes involved in mycolic acid synthesis may soon be available for detailed study.

The general pathway for mycolic acid biosynthesis has been elaborated mainly for the α-mycolates (Table 2.1) having two *cis*-cyclopropane rings (Figure 2.3). Many mycolic acids contain other very characteristic structural units (Table 2.1) whose biosynthetic pathways may offer further potential sites for drug attack. In a number of different mycolic acids cis double bonds are apparently transformed into trans double bonds with an adjacent methyl branch as well as the alternative conversion to a *cis*-cyclopropane ring (Figure 2.4). These trans unsaturations may then be transformed into trans cyclopropane rings (Figure 2.4), as discussed by Minnikin (1982). The biosynthetic interrelationships between the various oxygenated mycolic acids (Table 2.1) are beginning to emerge. Early studies (Etémadi and Gasche, 1965) indicated that the methyl branch next to the oxygen in wax-ester mycolates (Table 2.1) was derived from the methyl group of methionine (see Minnikin, 1982); no proposal for the biosynthetic incorporation of the methyl group was made. The characterization of epoxymycolates (Table 2.1) (Minnikin *et al.*, 1980; Daffé *et al.*, 1981; Minnikin, Minnikin and Goodfellow, 1982) has opened up a possible unified biosynthetic route for oxygenated mycolates (Figure 2.5). The *trans*-epoxide with an adjacent methyl branch (Minnikin, Minnikin and Goodfellow, 1982) could be formed directly from a methyl branched *trans*-alkene of the type discussed above (Figure 2.4). In support of this hypothesis, a mycolic acid having two *trans* unsaturations has been purified from *M. fortuitum* and *M. smegmatis* and assigned the tentative structure (**1**) (D.E. Minnikin, S.M. Minnikin and M. Goodfellow, unpublished results). This mycolic acid might be a direct precursor of an epoxymycolate (Table 2.1)

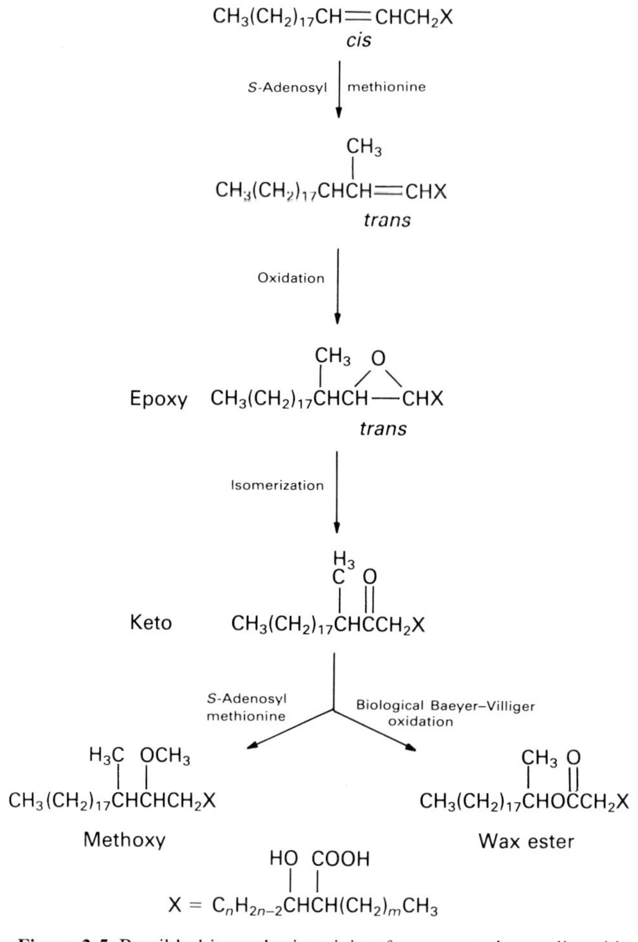


**Figure 2.5** Possible biosynthetic origin of oxygenated mycolic acids.

$$\underset{trans}{CH_3(CH_2)_lCH(CH_3)CH{=}}CH(CH_2)_{m-1}\underset{trans}{CH{=}}CHCH(CH_3)(CH_2)_nCH(OH)CH(COOH)(CH_2)_xCH_3$$

$l = 15, 17;\ m = 14, 16;\ n = 15, 17;\ x = 19, 21$

(**1**)

whose ring might be opened to yield a ketomycolate; a transformation which has chemical analogies (Minnikin, Minnikin and Goodfellow, 1982).

### *2.2.3 Outer membrane free lipids*

Lipids associated with the outer regions of the mycobacterial cell envelope have unusual structures and cover a very wide range of polarities. The most non-polar lipids are waxes such as the dimycocerosates of the phthiocerol family, being composed of multimethyl branched fatty acids esterified to long chain diols (Minnikin, 1982). At the other extreme are some very polar trehalose-based acylated oligosaccharides found, for example, in *M. kansasii* (Hunter *et al.*, 1983). The distribution of these complex lipids is discontinuous among mycobacterial species and is therefore of great value in classification and identification (Minnikin and Goodfellow, 1980; Dobson *et al.*, 1985). Since such complex lipids may be species, or even subspecies, specific, the pathways leading to their synthesis offer very precise targets for antimycobacterial drugs. In the following sections mycobacterial outer membrane free lipids will be introduced in order of increasing polarity.

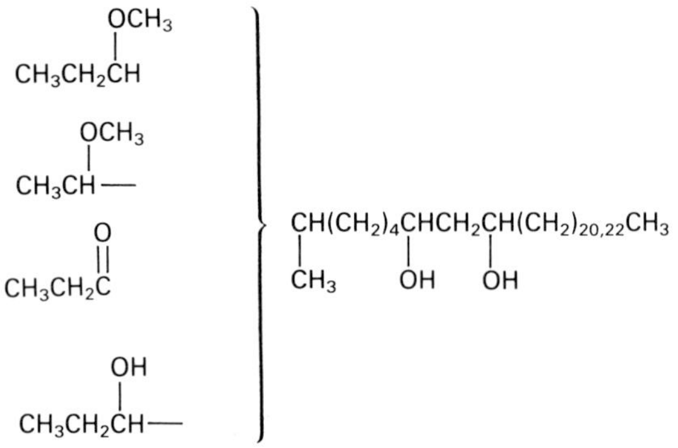

(**2**) $CH_3CH_2CH(OCH_3)$—
(**3**) $CH_3CH(OCH_3)$—
(**4**) $CH_3CH_2C(=O)$—
(**5**) $CH_3CH_2CH(OH)$—

} $CH(CH_3)(CH_2)_4CH(OH)CH_2CH(OH)(CH_2)_{20,22}CH_3$

$CH_3(CH_2)_{16-20}[CH_2CH(CH_3)]_{1-4}^{D}COOH$

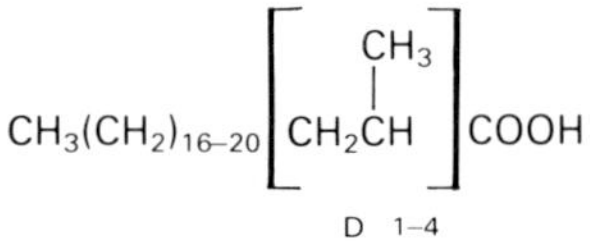

(**6**)

$CH_3CH_2CH(OCH_3)CH(CH_3)(CH_2)_4CH(OH)CH_2CH(OH)(CH_2)_n-C_6H_4-OH$

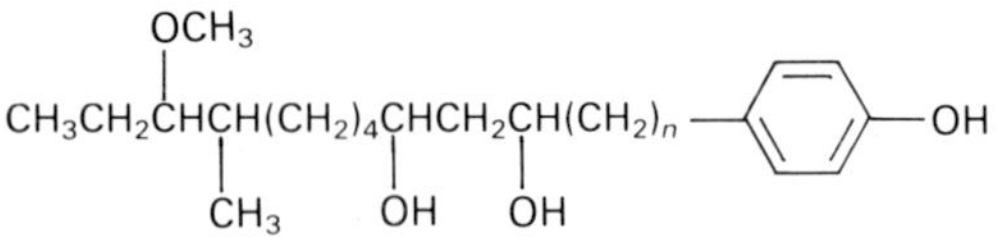

(**7**)

The phthiocerol family is composed of the related diols, phthiocerol A (**2**), phthiocerol B (**3**), phthiodiolone A (**4**) and phthiotriol A (**5**). The diol group is acylated by long chain multimethyl branched acids (**6**) which are termed mycocerosic acids (Minnikin, 1982). The components of the mycocerosates of the phthiocerol family vary in their detailed structures depending on the source, as summarized in Table 2.3. A very specific class of glycolipids is based on a closely related diol, phenolphthiocerol (**7**); the distribution of lipids based on phenolphthiocerol is shown in Table 2.4

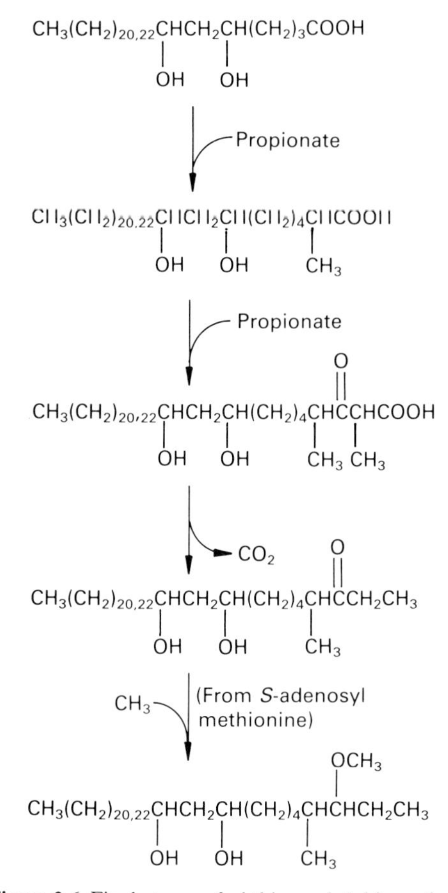


**Figure 2.6** Final stages of phthiocerol A biosynthesis.

The biosynthetic route leading to the phthiocerol A group is outlined in Figure 2.6 and, for phthiocerol B, Minnikin and Polgar (1965) suggested that acetate rather than propionate might be incorporated in the final stage. The most recent study on the biosynthesis of phenolphthiocerol (**7**) (Gastambide-Odier and Sarda, 1970) suggests that *p*-hydroxybenzoate, formed via shikimic and chorismic acids, is elongated to give a long chain diol. Early studies demonstrated the incorporation of pripionate into mycocerosic acids (**6**) (Gastambide-Odier, Delaumény and Federer, 1963), giving rise to the methyl branches. More recently a cell-free extract of *M. bovis* BCG showed that 18- and 20-carbon primers were elongated to give mycocerosates (Rainwater and Kolattukudy, 1983). The biosynthetic systems responsible for the glycosidation of phenolphthiocerol and the acylation of the diols (2–5, 7) with mycocerosate have not been studied.

Although *M. tuberculosis* synthesizes phthiocerol dimyocerosates (Table 2.3) and a dimycocerosate of the methyl ether of phenolphthiocerol (Table 2.4), glycolipids based on the latter diol are not produced (Minnikin, 1982). An

**Table 2.3** Dimycocerosates of the phthiocerol family

| Mycocerosic acid (**6**) | Species | Phthiocerol family (**2**, **3**) |
| --- | --- | --- |
| $C_{27}C_{32}$ | *M. bovis*[a] | $C_{34}$, $C_{36}$ |
| | *M. microti*[a] | |
| | *M. tuberculosis*[a] | |
| | *M. kansasii*[a,b] | $C_{27}$ |
| $C_{30}$–$C_{34}$ | *M. leprae*[c] | $C_{30}$, $C_{32}$ |
| $C_{27}$–$C_{30}$ | *M. marinum*[a] | $C_{28}$, $C_{30}$ |
| | *M. ulcerans*[b,d] | |

[a]Minnikin *et al.* (1985a).
[b]Diesters of phthiodiolone A (**4**) only; others have diesters of the diols (**2**, **3**).
[c]Draper *et al.* (1983), Hunter and Brennan (1983).
[d]Daffé *et al.* (1984); the multimethyl branched acids had the opposite stereochemistry to that of the mycocerosates (**6**). The stereochemistry of the acids from the lipids of the species, other than *M. tuberculosis*, requires study.

**Table 2.4** Lipids based on diacylated phenolphthiocerol

| Species | Substituent linked to phenol group |
| --- | --- |
| *M. bovis*[a] | 2-*O*-Methyl-*D*-rhamnose |
| *M. marinum*[a] | 3-*O*-Methyl-*L*-rhamnose |
| *M. kansasii*[a] | 2-*O*-Methyl-fucose, 2-*O*-Methyl-rhamnose, 2,4-Di-*O*-methyl-rhamnose |
| *M. leprae*[b] | 3-*O*-Methyl-rhamnose, 2,3-Di-*O*-methyl-rhamnose, 3,6-Di-*O*-methyl-glucose |
| *M. tuberculosis*[a] | Methyl |
| *M. ulcerans*[c] | None |

[a]See Minnikin (1982) for original references.
[b]See Hunter and Brennan (1983) which also contains details of two minor variants with modified oligosaccharides.
[c]Daffé *et al.* (1984).

alternative class of glycolipids, based on trehalose, has been characterized recently from *M. tuberculosis* (Minnikin *et al.*, 1985b). Both polar and non-polar types of these glycolipids are present and, as shown in Table 2.5, they vary in the number and structure of the acyl groups. The principal acyl groups are the well-known mycolipenic acids (**8**) (Minnikin, 1982) and C-24 and C-26 mycosanoic acids (**9**) (Cason, Lange and Urscheler, 1964). The absolute configuration of the chiral centres in these two classes of fatty acids is the opposite to that of the mycocerosic acids (**6**). It is possible that the mycosanoates (**9**) are

**Table 2.5** Trehalose mycolipenates from *M. tuberculosis* (Minnikin *et al.*, 1985b)

| Major acyl substituents | | Glycolipid[a] |
|---|---|---|
| Straight chain | Methyl branched | |
| Hexadecanoate, octadecanoate | $C_{17}$-mycolipenate (**8**) | A |
| | $C_{27}$-mycolipenate (**8**)<br>$C_{27}$-mycolipanolate[b] | B |
| | $C_{27}$-mycolipanolate[b]<br>$C_{24}$-mycosanoate (**9**) | C, D |

[a] A and B are a pair of relatively non-polar glycolipids, C and D being a more polar pair.
[b] $C_{27}$-mycolipanolic acid is 3-hydroxy-2,4,6-trimethyltetracosanoic acid.

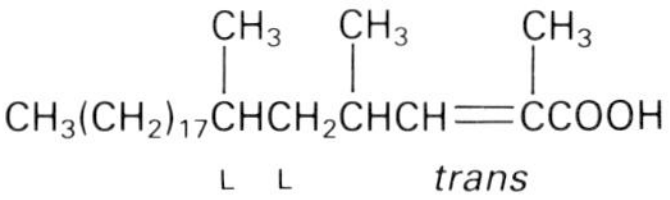

$$CH_3(CH_2)_{17}\underset{\text{L}}{\overset{CH_3}{\overset{|}{C}}}HCH_2\underset{\text{L}}{\overset{CH_3}{\overset{|}{C}}}HCH{=}\overset{CH_3}{\overset{|}{C}}COOH \quad (\textit{trans})$$

(**8**)

$$CH_3(CH_2)_{17}\underset{\text{L}}{\overset{CH_3}{\overset{|}{C}}}HCH_2\underset{\text{L}}{\overset{CH_3}{\overset{|}{C}}}HCOOH$$

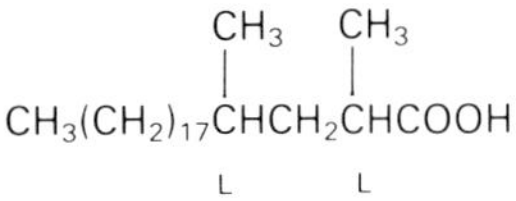

(**9**)

biosynthetic precursors of the mycolipenates (**8**), being elongated by addition of the elements of propionate (Gastambide-Odier, Delaumény and Kuntzel, 1966).

A family of multiacylated trehalose 2-sulphates have been characterized from *M. tuberculosis* H37Rv (Goren and Brennan, 1979; Minnikin, 1982; Goren, 1984). These sulpholipids have only been positively identified in strains of *M. tuberculosis*; suggestions of a ubiquitous distribution among mycobacteria (Khuller, Malik and Subramanyam, 1982; Malik *et al.*, 1982) have not been substantiated (Dhariwal, Dhariwal and Goren, 1984). The major sulpholipid from *M. tuberculosis* has the structure (**10**) and the essential structural features of the whole family are shown in Table 2.6. The principal acyl substituents are multimethyl branched acids termed phthioceranic (**11**) and hydroxyphthioceranic acids (**12**). The chirality of these acids is the same as that of the mycosanoates (**9**) and mycolipenates (**8**) but is the opposite to that of the mycocerosates. It appears that the biosynthesis of the phthioceranates (**11**) and hydroxyphthioceranates (**12**) may also involve the incorporation of propionate (Goren and Brennan, 1979).

The most celebrated complex mycobacterial lipids are the dimycolates of trehalose (**13**), the so-called 'cord-factors' (Asselineau and Asselineau, 1978b; Goren and Brennan, 1979; Minnikin, 1982). Trehalose dimycolates have been

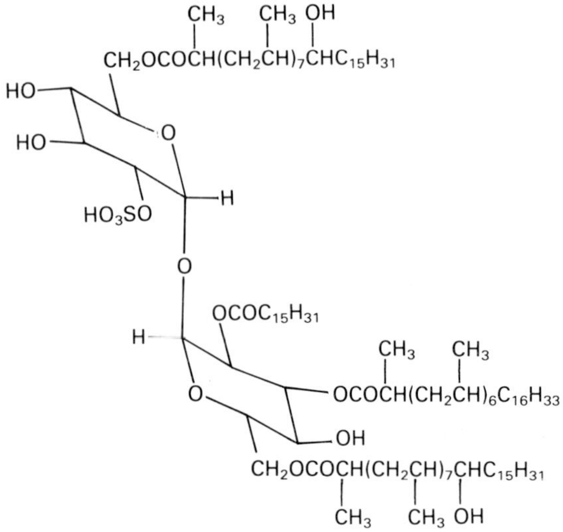

(**10**)

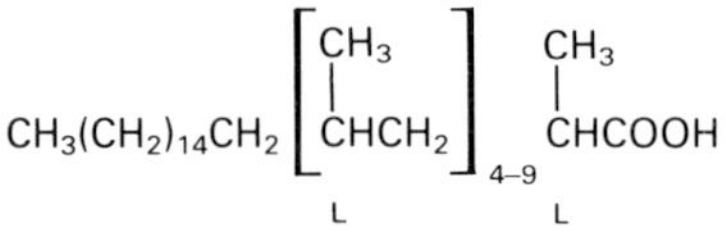

$CH_3(CH_2)_{14}CH_2[CH(CH_3)CH_2]_{4-9}CH(CH_3)COOH$ (L, L)

(**11**)

$CH_3(CH_2)_{14}CH(OH)[CH(CH_3)CH_2]_{4-9}CH(CH_3)COOH$ (L, L)

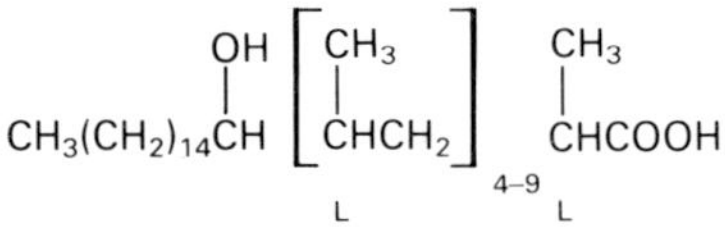

(**12**)

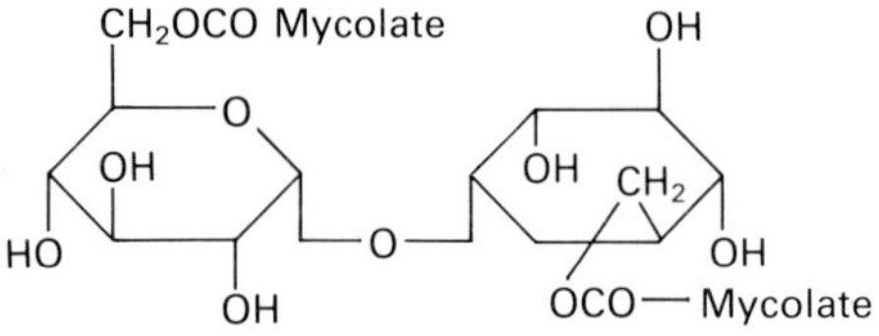

(**13**)

Fatty acyl-Phe-*allo*-Thr-Ala-alaninol-sugar
|
O
|
Basal sugar unit–specific sugar(s)
} (Acetyl)$_n$

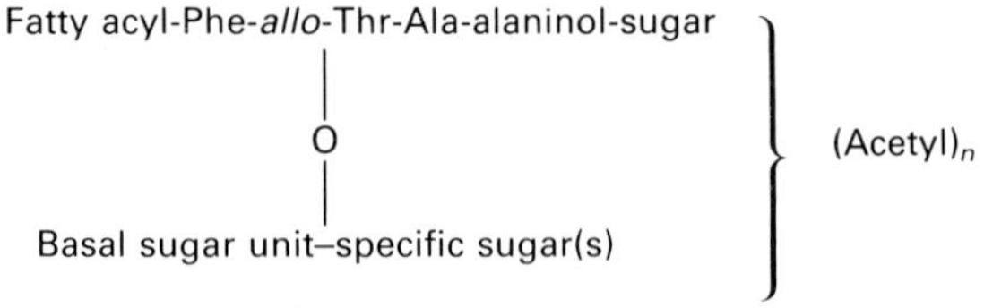

(**14**)

found in practically all mycobacteria investigated for their presence but, since the amounts detected are very variable, it is considered that their main role may be as biosynthetic intermediates rather than as structural components (Minnikin, 1982). Cord-factors do have some toxicity and have antitumour activity when used in conjunction with mycobacterial wall components (Goren and Brennan, 1979).

Glycosylated peptidolipids are very characteristic components of the complex

**Table 2.6** Sulphatides of *M. tuberculosis* H37Rv

| Lipid[a] | Trehalose substitution | Palmitate or stearate | Phthioceranate (**11**) | Hydroxyphthioceranate (**12**) |
|---|---|---|---|---|
| SL - II′ | 2,4,6,6′ | 1 | 0 | 3 |
| SL - II | 2,3,6,6′ | 1 | 0 | 3 |
| SL - I(**10**) | 2,3,6,6′ | 1 | 1 | 2 |
| SL - I′ | 2,3,6,6′ | 1 | 2 | 1 |
| SL - III | 2,3,6 | 1 | 0 | 2 |

[a]See Goren (1984) for explanation of abbreviations.

free lipids of a number of mycobacterial species. All these glycolipids appear to be based on an acylated tetrapeptide (**14**) and representative structures are shown in Table 2.7. Both polar and apolar classes of glycopeptidolipids occur; the former group are lipid antigens having serotype-specific sugars (Goren and Brennan, 1979; Brennan, 1984). Little is known about the biosynthesis of glycopeptidolipids but a reductase which converts L-alanine to L-alaninol has been identified in *M. avium* (Bruneteau and Michel, 1971).

The above glycolipids are readily recognized by their relative stability to alkali (Brennan, Heiferts and Ullom, 1982) but a completely distinct group of alkali-labile lipooligosaccharides have recently been identified (Hunter *et al.*, 1983; Tsang *et al.*, 1984). Characteristic thin-layer chromatographic patterns of these oligosaccharides have been recorded for representatives of *M. fortuitum*, *M. gordonae*, *M. szulgai*, *M. kansasii* and *M. marinum* (Tsang *et al.*, 1984), it being notable that the latter two species also produce glycosylphenolphthiocerols (Table 2.4). Structural studies have only been carried out on the antigenic lipooligosaccharides from *M. kansasii* (Hunter *et al.*, 1983) and one of the partially identified structures is shown (**15**). It is interesting to note that the long acyl substituent of one of the lipooligosaccharides from *M. kansasii* was a single fatty acid, 2,4-dimethyltetradecanoic acid, accompanied by acetyl groups (Hunter *et al.*, 1983). Two pyruvylated trehalose-based glycolipids have been characterized from *M. smegmatis* (Saadat and Ballou, 1983).

A family of iron-binding compounds, the mycobactins, have been characterized from all species of mycobacteria so far examined, excepting *M. paratuberculosis* (Ratledge, 1982). These compounds chelate iron and facilitate its transport through the cell envelope; the structure of the mycobactin from *M. tuberculosis* is shown (**16**). Other species of *Mycobacterium* produce related mycobactins whose structural variations are of value in classification and identification (Ratledge, 1982). The detailed biosynthetic pathways leading to the mycobactins have not been elucidated but the aromatic nucleus is formed from salicylic acid which originates through the shikimic acid pathway (Ratledge, 1982).

### 2.2.4 *Plasma membrane lipids*

In comparison with the outer membrane free lipids, mycobacterial plasma

**Table 2.7** Essential structures of mycobacterial glycopeptidolipids (Brennan, 1984)

| Lipid type | Serovar | Sugars | | |
|---|---|---|---|---|
| | | *Allo*-threonine-linked | | Alaninol-linked |
| | | Base unit | Specific oligosaccharide | |
| Apolar | All | 6-*d*-Tal or 3-Me-6-*d*-Tal[a] | — | 3,4-$Me_2$-Rha[a] or 3-Me-Rha |
| Polar | 8 | α-L-Rha (1 → 2)-6-L-*d*-Tal | 4,6-(1′-carboxyethylidene)-3-Me-β-D-Glc (1 → 3)- | 3,4-$Me_2$-Rha |
| | 9 | α-L-Rha (1 → 2)-6-L-*d*-Tal | 2,3-$Me_2$-α-L-Fuc-(1 → 4)-2,3-$Me_2$-α-L-Fuc-(1 → 3)- | 3,4-$Me_2$-Rha |
| | 25 | α-L-Rha (1 → 2)-6-L-*d*-Tal | 2-Me-α-L-Fuc (1 → 4)-2-Me-α-L-Fuc-(1 → 3)- | 3,4-$Me_2$-Rha |

[a]Abbreviations exemplified by: 3-Me-6-*d*-Tal, 3-*O*-methyl-6-deoxytalose; 3,4-$Me_2$-Rha, 3,4-di-*O*-methylrhamnose.

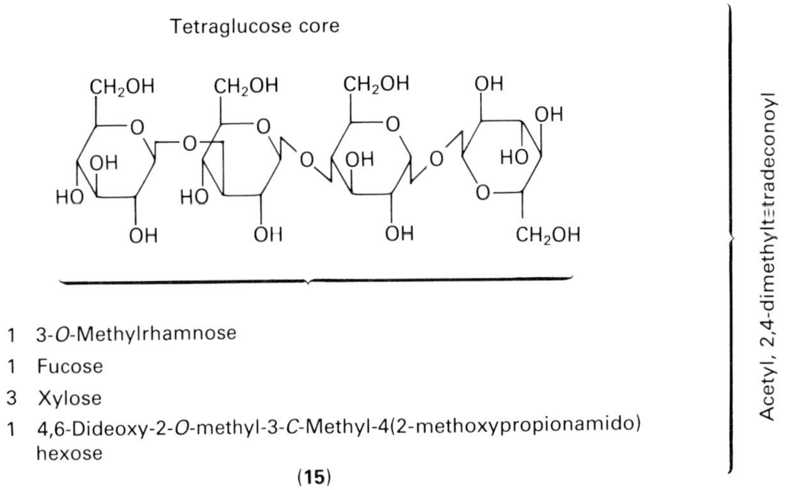
Tetraglucose core
CH2OH
CH2OH
CH2OH
OH
OH
HO
HO
HO
CH2OH
1 3-O-Methylrhamnose
1 Fucose
3 Xylose
1 4,6-Dideoxy-2-O-methyl-3-C-Methyl-4(2-methoxypropionamido) hexose
Acetyl, 2,4-dimethyltetradeconoyl

(**15**)

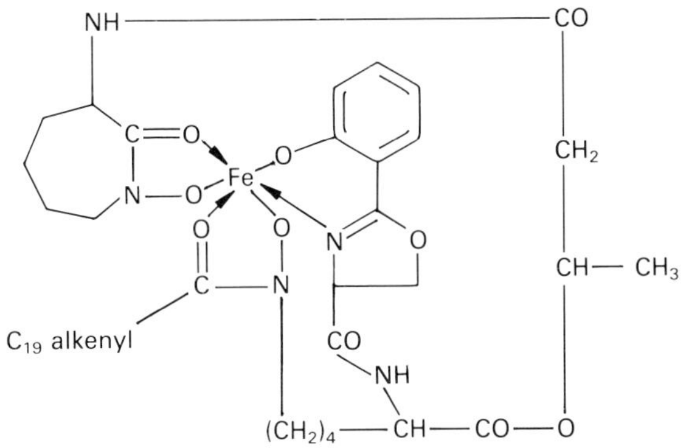
NH
CO
CH2
CH— CH3
Fe
C19 alkenyl
CO
NH
(CH2)4—CH—CO—O

(**16**)

membrane free lipids are mainly conventional in character. The structures of the principal mycobacterial phospholipids are given in Table 2.8; acylated ornithine amides have also been isolated from a strain of *M. bovis* (Promé, Lacave and Lanéelle, 1969). Phosphatidylinositol mannosides were originally characterized from mycobacteria (Minnikin, 1982) but they are also present in many other actinomycetes (Minnikin and O'Donnell, 1984). The fatty acid components of the membrane polar lipids are similar to those of many other actinomycetes, consisting of mixtures of straight chain, monounsaturated and 10-methyl branched acids (Minnikin and O'Donnell, 1984). Menaquinones and carotenoid pigments are also components of mycobacterial plasma membranes; their structures are not unusual (Minnikin, 1982).

### *2.2.5 Chemical model of the mycobacterial envelope*

The innermost layer in the mycobacterial envelope is the plasma membrane

**Table 2.8** Structures of the principal mycobacterial polar lipids

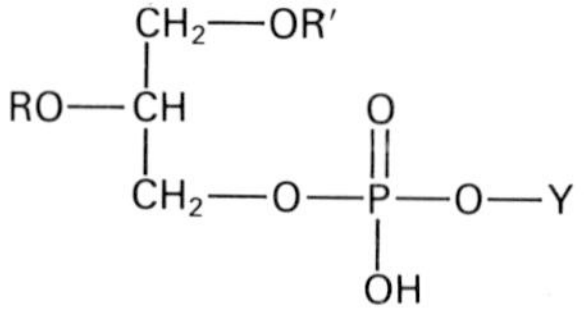

$CH_2$—OR′
RO—CH
$CH_2$—O—P(=O)(OH)—O—Y

R, R′ = long chain acyl

Y = ethanolamine, phosphatidylethanolamine; Y = glycerol, phosphatidylglycerol (PG); Y = PG, diphosphatidylglycerol; Y = inositol, phosphatidylinositol

$CH_2$—OR′
RO—CH
$CH_2$—O—P(=O)(OH)—O—[inositol ring: OMan, OH (R″, R‴); OH; OH; O(Man)$_n$]

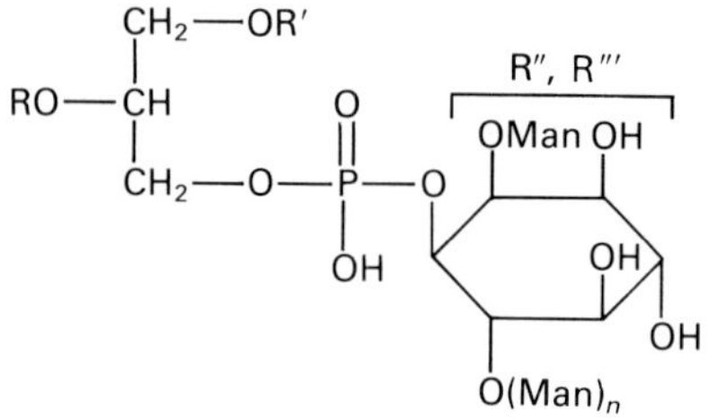

Man = mannose

Phosphatidylinositol mannosides:
R, R′, R″ = acyl; R‴ absent; $n$ = 1: monoacylphosphatidylinositol dimannoside
R, R′, R″, R‴ = acyl; $n$ = 1: diacylphosphatidylinositol dimannoside
R, R′, R″ = acyl; R‴ absent; $n$ = 4: monoacylphosphatidylinositol pentamannoside
R, R′, R″, R‴ = acyl; $n$ = 4: diacylphosphatidylinositol pentamannoside

whose composition and organization is not apparently different from that of such membranes in other bacteria, as noted in the previous section. Plasma membranes are considered to be based on an essential lipid bilayer, interacting with proteins, which forms the permeability barrier enclosing the cytoplasm. It is also the site of much of the essential enzymic activity in the cell envelope; the biosynthesis of many of the external components of the envelope may be at least initiated therein. It is probable that, as in the majority of bacteria studied, the organelle which encloses the plasma membrane is the peptidoglycan. This rigid sacculus supports the flexible plasma membrane and its composition (Figure 2.1) is similar to peptidoglycans from other bacteria. An unusual feature, shared with other actinomycetes (Minnikin and O'Donnell, 1984), is the presence of *N*-glycollated muramic acids. Arabinogalactan polysaccharides are also characteristic components of all actinomycetes which have mycolic acids in their envelopes (Minnikin and O'Donnell, 1984).

Mycobacteria differ from other mycolic acid-containing actinomycetes in the complexity of the types of mycolic acids and associated free lipids. All of these lipids are based on very unusual long chain components, structurally very different to those which form the hydrophobic sections of the plasma membrane amphipathic polar lipids. It is considered that fatty acid chains interact together in a cooperative manner to produce a membrane hydrophobic core having suitable physical properties. Indeed, the proportions of the different bacterial fatty acid types may vary with temperture to maintain membrane fluidity within certain limits. The mycolic acid chain lengths in *M. phlei* have been shown also to vary with temperature, strongly suggesting that such acids participate in a membraneous organelle whose physical properties are important in cell integrity.

Considerations of how all the diverse free lipid structures might interact with covalently bound mycolic acids resulted in the proposal of a chemical model of the lipid domains in the mycobacterial envelope (Figure 2.7) (Minnikin, 1982). The basis of the proposed arrangement is a monolayer of mycolic acids bound to arabinogalactan and arranged with their main and branched chains parallel, producing a relatively close-packed inner structural permeability barrier (Figure 2.7). The regular spacing of the unsaturations and oxygen functions in mycolic acids (Table 2.1) results in the two additional lipid domains shown in Figure 2.7. The so-called parallel binding region (Figure 2.7) may offer hydrophobic anchoring sites for the long acyl components of the complex free lipids with the outer hydrophobic interaction region providing an intermediate zone bordering on the hydrophilic exterior (Figure 2.7).

The acyl chains of the complex free lipids are very different to those of the amphipathic polar lipids found in plasma membranes. Waxes based on phthiocerol and the glycosyl phenolphthiocerols contain mycocerosic acids (**6**) whose structures comprise a long straight chain portion attached to a multimethyl branched section. Such structures have been rationalized by the suggestion (Minnikin, 1982) that in such acyl components the long straight chain unit may be inserted into the parallel binding region of the mycolate matrix (Figure 2.7). The multimethyl branched part of the fatty acid is considered to be more flexible and may interact in a less specific manner with the terminal sections of the mycolate chains. The length of the methyl branched section may indeed determine the depth of insertion, acting like a 'key' being inserted into the mycolate matrix. For example, the phthioceranates (**11**, **12**), having up to nine methyl branches, may be designed to project the hydrophilic sulphated trehalose units in the sulpholipids (Table 2.5) beyond the limits of the mycolate chains (Figure 2.7). The smaller number of methyl branches in the mycocerosates (**6**) would allow the dimycocerosates of phthiocerol and phenolphthiocerol to be locked in more tightly, with only the sugars of the glycolipids reaching to the surface.

A number of relatively shorter chain methyl branched acids have been identified from several mycobacteria (Tisdall, Roberts and Anhalt, 1979; Julák, Tureček and Miková, 1980). A typical example is the 2,4-dimethyltetradecanoate from the lipooligosaccharides of *M. kansasii* (**15**). The precise structures of these lipids and the number of acyl substituents are not yet known but it is possible that the methyl branched acids are distributed over the tetraglucose core, offering multipoint lipophilic anchors to bind these surface antigens to the mycolic acid matrix (Figure 2.7). The wider spacing of the lipid anchors may allow effective binding with alkyl chains shorter than those found in the mycocerosates (**6**) and

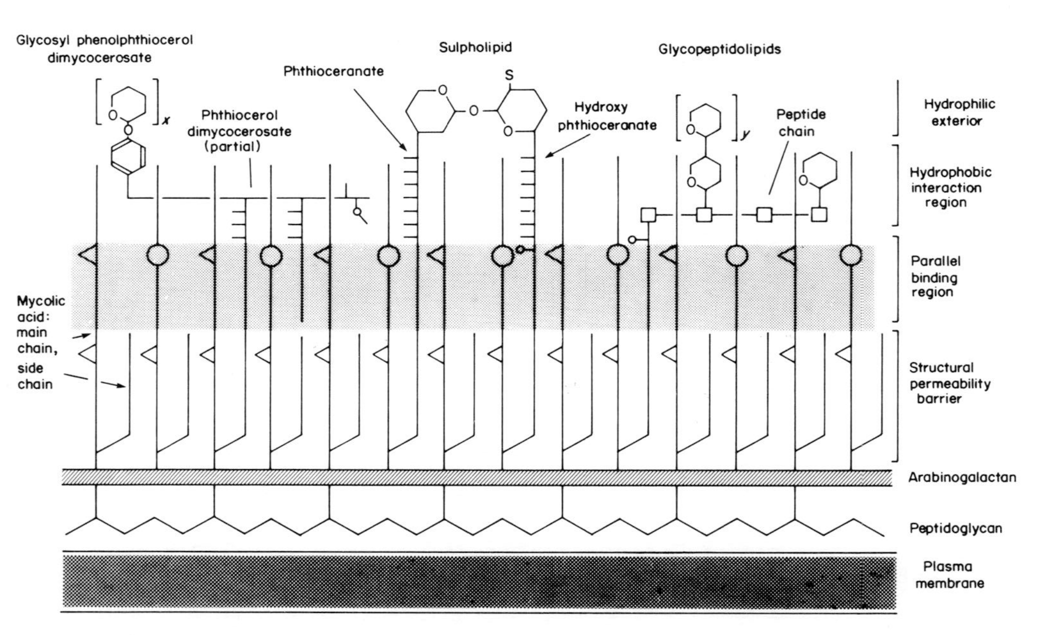


**Figure 2.7** Chemical model of lipid interactions in mycobacterial envelopes, modified from Minnikin (1982). The whole model is not drawn to scale but the lengths of the hydrocarbon chains are approximated. Lateral spikes indicate methyl branches in the acyl chains of lipids. Mycolic acids are considered to pack together with the side chains parallel to the main chain and unsaturations are indicated by triangles and oxygen functions by circles. Lipids such as the dimycocerosates of phthiocerol A (Table 2.3), glycosyl phenolphthiocerol (Table 2.4) and sulphated trehalose phthiocernates (Table 2.6) may be inserted as shown, with the methyl branched regions of their acyl chains limiting the depth of insertion into the parallel binding region of the mycolic acid matrix. The characteristic sugars *x* in the glycosyl phenolphthiocerol dimycocerosates (Table 2.4) may then be accessible on the cell surface as would be the hydrophilic portion of the sulpholipids. The phthiocerol dimycocerosate (Table 2.3) is represented in partial form for simplicity; it essentially lacks the glycosyl phenol portion and may be regarded as an 'inert filler' in comparison with the 'active' glycolipid. Similarly, the polar glycopeptidolipids (Table 2.7) having serotype-specific oligosaccharides (*y*) are active variants of the non-polar types (Table 2.7). Anchorage of the glycopeptidolipids may be aided by hydrogen bonding between oxygen functions in the acyl chains and those in the mycolic acid main chains. The hydroxyl group in a hydroxyphthioceranic acid (**12**) substituent of a sulpholipid may also provide an opportunity for hydrogen bonding. Since the model is a two-dimensional representation, the structures of the complex free lipids have not been drawn completely. It should also be noted that the lipids used as examples do not all occur together in the same organism.

phthioceranates (**11**, **12**). Acetyl groups on the oligosaccharides (**15**) may serve to adjust the polarity to a level suitable for interaction with the terminal sections of the mycolate chains (Figure 2.7). The more unusual sugars (**15**) may possibly project to the hydrophilic exterior to express their biological functions.

The long chain components of glycopeptidolipids (Table 2.7) conform to a different pattern, having a long straight chain with usually an oxygen function near the carboxyl group. For these lipids, the depth of insertion into the outer membrane matrix may be governed by interactions between oxygen functions in the glycopeptidolipid acyl chains and those in the main chains of the oxygenated mycolic acids (Figure 2.7). The indications are, therefore, that complex mycobacterial free lipids may be designed to fit together specifically with the mycolic acid matrix, details of the interactions varying among the different species.

It is apparent that the complex free lipids fall into two functional groups which may be regarded as inert 'filler' lipids and biologically active types (Minnikin, 1982). Examples of the former group include the phthiocerol dimycocerosates (Table 2.3), apolar types of trehalose mycolipenates (Table 2.5) and apolar glycopeptidolipids (Table 2.7). These lipids may provide an inert defensive bastion among which the more active offensive types are interspersed. The active types are the proven surface antigens such as the glycosyl phenolphthiocerols (Table 2.4), polar glycopeptidolipids (Table 2.7), sulpholipids (Table 2.6) and lipooligosaccharides of the type found in *M. kansasii* (**15**). The polar varieties of the trehalose mycolipenates from *M. tuberculosis* (Table 2.5) may eventually be added to the latter class.

## 2.3 Chemical targets

The chemical components of the envelopes of mycobacteria range very widely in structure but a unified picture is emerging concerning their mutual interactions. It is also becoming more clear exactly which lipids are of prime importance in the structure and function of an effective surface membrane organelle. Such knowledge is enabling the most vital biosynthetic pathways to be pinpointed and indicates links in these routes which may be attacked by antibiotics. Chemical targets in mycobacterial envelopes may be considered at several different levels of selectivity, varying from subspecies serotype specificity to components widespread in many actinomycete taxa. Possible strata of selectivity are proposed in Table 2.9 and each class is amplified in the following sections.

The highest degree of specificity is shown by the serotype-specific surface lipid antigens, such as the polar glycopeptidolipids (Table 2.7) characteristic of the *M. avium*, *M. intracellulare* and *M. scrofulaceum* serocomplex. Antibiotics which could inhibit the incorporation of particular antigenic determinant sugars into these lipids would be selective for single serotypes. Such high selectivity might be of value in controlling a serovar with exceptional virulence. Representatives of *M. gordonae* also appear to produce strain-dependent variations in patterns of

**Table 2.9** Levels of selectivity in chemical targets for drugs in mycobacterial cell envelopes

| Specificity | Selective targets |
|---|---|
| Subspecies | Glycopeptidolipids (Table 2.7), lipooligosaccharides (**15**) |
| Species | Glycosylphenolphthiocerols (Table 2.4), trehalose mycolipenates (Table 2.5), sulpholipids (Table 2.6), mycobactins (**16**) |
| Species groups | All dimyocerosates of phthiocerol family (Table 2.3) and glycosylphenolphthiocerols (Table 2.4). Different mycolic acid types |
| Genus | General mycobacterial mycolate synthesis and incorporation into arabinogalactan. Mycobactins (**16**) |
| Mycolate-containing genera | Synthesis and assembly of arabinogalactan mycolate organelle |
| Certain actinomycetes | Tuberculostearic acid, phosphatidylinositol mannosides |

lipooligosaccharides (Tsang *et al.*, 1984), so similar discriminatory potential may also be possible within this species.

Other characteristic lipids are found in all representatives of particular species and antibiotics targeted at their biosynthetic pathways would be species specific. The best examples are the glycosyl phenolphthiocerols (Table 2.4) whose distribution is limited to *M. bovis*, *M. kansasii*, *M. leprae* and *M. marinum*. Inhibition of the incorporation of the determinant sugars would discriminate between these species. The glycosyl phenolphthiocerol from *M. leprae* (Table 2.4), as well as being a component of the outer membrane (Figure 2.7) (Minnikin, 1982), is also exported into the region surrounding the organism, contributing to a transparent zone seen in electron micrographs (Draper, 1982). Indeed, in a recent study (Mehra *et al.*, 1984) it has been shown that these extracellular lipids induce suppression of mitogenic responses of lymphocytes from lepromatous patients. *Mycobacterium tuberculosis* might be selectively attacked by targeting antibiotics at the trehalose mycolipenates (Table 2.5) or the sulpholipids (Table 2.6) and their long chain components (**8**, **9**, **11**, **12**). The structures of mycobactins (**16**) appear to be characteristic of particular species (Ratedge, 1982), offering more species-specific targets.

Groups of mycobacterial species might be attacked in several different ways by targeting antibiotics at common biosynthetic pathways. For example, all mycobacteria which synthesize dimycocerosates of the phthiocerol family (Table 2.3), glycosyl phenolphthiocerol (Table 2.4) and glycopeptidolipids (Table 2.7) presumably do so by a common route. Similarly, organisms producing different mycolic acid types (Table 2.1) could be attacked as a group, those synthesizing epoxymycolates for example.

Since mycobacterial mycolic acids are distinct in structure from those of other mycolic acid-containing taxa, it would be expected that the pathways leading to the basic structures of such acids might be similarly susceptible to drug attack. Isoniazid is considered to inhibit mycolate synthesis (Takayama & Qureshi, 1984) but it has a wide range of activity against different mycobacteria, suggesting either that its action may not be very specific or that its access to a particular site is differentially hindered. Mechanisms for the transfer of mycobacterial mycolic acid units to the arabinogalactan (Figure 2.2) may be different from those in other mycolic acid-containing actinomycetes, offering another site limited to the genus *Mycobacterium*. Particular attention should be paid to the possibility of aiming antibiotics against the general pathways leading to mycobactins, such as that from *M. tuberculosis* (**16**). These latter compounds are vital for the uptake of essential iron (Ratledge, 1982) and offer relatively accessible effective targets.

It is possible that some of the biochemical pathways for cell envelope components are shared between mycobacteria and the other genera with an outer membrane based on a peptidoglycan-linked arabinogalactan–mycolate matrix, representatives of the genera *Corynebacterium*, *Nocardia* and *Rhodococcus* being good examples (Minnikin and O'Donnell, 1984). In particular, the arabinogalactan organelle (Figure 2.2) must be assembled in a very precise manner and the pathways involved offer good targets. As noted above, ethambutol apparently interferes with arabinogalactan assembly (Takayama and Kilburn, 1984), and this is a good illustration of the potential of such agents which inhibit the assembly of envelope components. It is not clear, however, that the arabinogalactans from different genera are identical, there being a lack of accurate systematic data. It is also possible that some aspects of mycolate biosynthesis are common to all these organisms, the condensation which introduces the chain in the 2-position for example (Figure 2.3). The great range of mycolate types would require, however, wide flexibility in any common enzyme systems.

Mycobacterial lipids, such as tuberculostearic acid and the phosphatidylinositol mannosides (Table 2.8) are found in a wide range of actinomycetes (Minnikin and O'Donnell, 1984) and would not provide very specific targets. The phosphatidylinositol mannosides appear to be essential components of the mycolic acid-containing and certain other actinomycetes; drugs inhibiting their synthesis would have a broad spectrum of activity.

The rich variety of unusual chemical components present in the cell envelopes of mycobacteria offer, therefore, a range of targets having varying specificity. Access to such targets is very difficult at present because so little is known about the detailed biochemistry of these compounds. It is currently very difficult, therefore, to design drugs against specific enzymes involved in mycobacterial envelope biochemical processes. The slow-growing habit and cell complexity of mycobacteria greatly hinder biochemical studies and an enormous amount of work would be necessary to unravel the enzymology of even a single envelope component. The main practical approach at present would appear to be the full elucidation of the modes of action of established antimycobacterial agents and the development of modified drugs having enhanced potency. It is important to note, however, that when organisms develop resistance to particular drugs, it is unlikely that slightly modified analogues of such drugs would have the capacity to overcome this resistance.

## 2.4 Acknowledgements

Studies on the subject of the present paper were carried out in collaboration with M. Goodfellow, G. Dobson, J.H. Parlett, S.M. Minnikin, P. Draper, M. Ridell, M. Magnusson and F. Portaels. Research grants were provided by the Medical Research Council (G974/522/S; G8216538), Science and Engineering Research Council (GRA 88651), IMMLEP Steering Committee of the UNDP/World Bank/WHO Special Programme for Research and Training in Tropical Diseases (T16/1818/L4/29) and the British Leprosy Relief Association.

## 2.5 References

1. R.J. Anderson, (1941). *Chem. Rev.*, **29**, 225–243.
2. C. Asselineau, and J. Asselineau, (1978a). *Ann. Microbiol.*, **129A**, 49–69.
3. C. Asselineau, and J. Asselineau, (1978b). *Prog. Chem. Fats Other Lipids*, **16**, 59–99.
4. J. Asselineau, (1966). *The Bacterial Lipids*, Hermann, Paris.
5. J.H. Bates, (1984). In *The Mycobacteria: A Sourcebook* (Eds G.P. Kubica and L.G. Wayne), Part B, pp. 991–1005, Marcel Dekker, New York.
6. P.J. Brennan, (1984). In *The Mycobacteria: A Sourcebook* (Eds) G.P. Kubica and L.G. Wayne), Part A, pp. 467–489, Marcel Dekker, New York.
7. P.J. Brennan, M. Heiferts, and B.P. Ullom, (1982). *J. Clin. Microbiol.*, **15**, 447–455.
8. M. Bruneteau, and G. Michel, (1971). *FEBS Lett.*, **14**, 57–60.
9. J. Cason, G.L. Lange, and H.R. Urscheler, (1964). *Tetrahedron*, **20**, 1955–1961.
10. M.H. Cynamon, and G.S. Palmer, (1983). *Antimicrob. Agents Chemother.*, **24**, 429–431.
11. M. Daffé, M.A. Lanéelle, C. Asselineau, V. Lévy-Frébault, and H. David, (1983). *Ann. Microbiol.*, **134B**, 241–256.
12. M. Daffé, M.A. Lanéelle, G. Puzo, C. Asselineau, (1981). *Tetrahedron Lett.*, **22**, 4515–4516.
13. M. Daffé, M.A. Lanéelle, J. Roussel, and C. Asselineau, (1984). *Ann. Microbiol.*, **135A**, 191–201.
14. K.R. Dhariwal, G. Dhariwal, and M.B. Goren, (1984). *Am. Rev. Respir. Dis.*, **130**, 641–646.
15. G. Dobson, D.E. Minnikin, S.M. Minnikin, J.H. Parlett, M. Goodfellow, M. Ridell, and M. Magnusson, (1985). In *Chemical Methods in Bacterial Systematics* (Eds. M. Goodfellow and D.E. Minnikin), pp. 237–265, Academic Press, London.
16. P. Draper, (1982). In *The Biology of the Mycobacteria* (Eds. C. Ratledge and J.L. Stanford), pp. 9–52, Academic Press, London.
17. P. Draper, S.N. Payne, G. Dobson, and D.E. Minnikin, (1983). *J. Gen. Microbiol.*, **129**, 859–863.
18. A.H. Etémadi, (1967). *Bull. Soc. Chim. Biol.*, **49**, 695–706.
19. A.H. Etémadi, and J. Gasche, (1965). *Bull. Soc. Chim. Biol.*, **47**, 2095–2104.
20. M. Gastambide-Odier, J.M. Delaumény, and H. Kuntzel, (1966). *Biochim. Biophys. Acta*, **125**, 33–42.
21. M. Gastambide-Odier, J.M. Delaumény, and E. Lederer, (1963). *Chem. Ind.*, 1285–1286.
22. M. Gastambide-Odier, and E. Lederer, (1960). *Biochem. Zeitschr.*, **333**, 285–295.
23. M. Gastambide-Odier, and P. Sarda, (1970). *Pneumonologie*, **142**, 241–255.
24. C.R. Goucher, and M.C. Cabot, (1981). *Lipids*, **16**, 146–148.
25. T. Godal, and L. Levy, (1984). In *The Mycobacteria: A Sourcebook* (Eds. G.P. Kubica and L.G. Wayne), Part B, pp. 1083–1128, Marcel Dekker, New York.
26. M.B. Goren, (1984). In *The Mycobacteria: A Sourcebook* (Eds. G.P. Kubica and L.G. Wayne), Part A, pp. 379–415, Marcel Dekker, New York.
27. M.B. Goren, and P.J. Brennan, (1979). In *Tuberculosis* (Ed. G.P. Youmans), pp. 63–193, W. B. Saunders, Philadelphia.
28. J.E. Hawkins, (1984). In *The Mycobacteria: A Sourcebook* (Eds. G.P. Kubica and L.G. Wayne), Part A, pp. 177–193, Marcel Dekker, New York.

29. M. Hooper, (1985). *Lepr. Rev.*, **56**, 57–60.
30. M. Hooper, and M.G. Purohit, (1983). In *Progress in Medicinal Chemistry* (Eds. G.P. Ellis and G.B. West), Vol. 20, pp. 1–81, Elsevier, Amsterdam.
31. S.W. Hunter, and P.J. Brennan, (1983). *J. Biol. Chem.*, **258**, 7556–7562.
32. S.W. Hunter, R.C. Murphy, K. Clay, M.B. Goren, and P.J. Brennan, (1983). *J. Biol. Chem.*, **258**, 10481–10487.
33. J. Julák, F. Tureček, and Z. Miková, (1980). *J. Chromatogr.*, **190**, 183–187.
34. G.K. Khuller, U. Malik, and D. Subramanyam, (1982). *Tubercle*, **63**, 107–111.
35. J.O. Kilburn, and K. Takayama, (1981). *Antimicrob. Agents Chemother.*, **20**, 401–404.
36. V.M. Kulkarni, and J.K. Seydel, (1983). *Chemotherapy*, **29**, 58–67.
37. E. Lederer, A Adam, R. Ciorbaru, J.-F. Petit, and J. Wietzerbin, (1975). *Mol. Cell. Biochem.*, **7**, 87–104.
38. L. Levy, (1975). *Am. Rev. Respir. Dis.*, **111**, 703–705.
39. U. Malik, A.V. Prabhudesai, G.K. Khuller, and D. Subrahmanyam, (1982). *IRCS Med. Sci.*, **10**, 910.
40. V. Mehra, P.J. Brennan, E. Rada, J. Convit, and B.R. Bloom, (1984). *Nature*, **308**, 194–196.
41. D.E. Minnikin, (1982). In *The Biology of the Mycobacteria* (Eds. C. Ratledge and J.L. Stanford), pp. 95–184, Academic Press, London.
42. D.E. Minnikin, G. Dobson, M. Goodfellow, M. Magnusson, and M. Ridell, (1985a). *J. Gen. Microbiol.*, **131**, 1375–1381.
43. D.E. Minnikin, G. Dobson, D. Sesardic, and M. Ridell, (1985b). *J. Gen. Microbiol.*, **131**, 1369–1374.
44. D.E. Minnikin, and M. Goodfellow, (1980). In *Microbiological Classification and Identification* (Eds. M. Goodfellow and R.G. Board), pp. 189–256, Academic Press, London.
45. D.E. Minnikin, I.G. Hutchinson, A.B. Caldicott, and M. Goodfellow, (1980). *J. Chromatogr.*, **188**, 221–233.
46. D.E. Minnikin, S.M. Minnikin, and M. Goodfellow, (1982). *Biochim. Biophys. Acta*, **712**, 616–620.
47. D.E. Minnikin, S.M. Minnikin, J.H. Parlett, and M. Goodfellow, (1985c). *Zbl. Bakt. Hyg.*, **A**, **259**, 446–460.
48. D.E. Minnikin, S.M. Minnikin, J.H. Parlett, M. Goodfellow, and M. Magnusson, (1984). *Arch. Microbiol.*, **139**, 225–231.
49. D.E. Minnikin, and A.G. O'Donnell, (1984). In *The Biology of the Actinomycetes* (Eds. M. Goodfellow, M. Mordarski and S.T. Williams), pp. 337–388, Academic Press, London.
50. D.E. Minnikin, and N. Polgar, (1965). *Chem. Comm.*, **1965**, 495.
51. A. Misaki, N. Seto, and I. Azuma, (1974). *J. Biochem.*, **76**, 15–27.
52. S.R. Pattyn, (1984). In *The Mycobacteria: A Sourcebook* (Eds. G.P. Kubica and L.G. Wayne), Part B, pp. 1129–1135, Marcel Dekker, New York.
53. J.C. Promé, C. Lacave, and M.A. Lanéelle, (1969). C.R. Acad. Sci., **269C**, 1664–1667.
54. N. Qureshi, N. Sathyamoorthy, and K. Takayama, (1984). *J. Bacteriol.*, **157**, 46–52.
55. D.L. Rainwater, and P.E. Kolattukudy, (1983). *J. Biol. Chem.*, **258**, 2979–2985.
56. J.W. Raleigh, (1984). In *The Mycobacteria: A Sourcebook* (Eds. G.P. Kubica and L.G. Wayne), Part B, pp. 1007–1020, Marcel Dekker, New York.
57. C. Ratledge, (1982). In *The Biology of the Mycobacteria* (Eds. C. Ratledge and J.L. Stanford), pp. 185–271, Academic Press, London.
58. S. Saadat, and C.E. Ballou, (1983). *J. Biol. Chem.*, **258**, 1813–1818.
59. J. Sarracent, and C.M. Finlay, (1984). *Int. J. Lepr.*, **52**, 154–158.
60. K. Takayama, and E.L. Armstrong, K.A. Kunugi, and J.O. Kilburn, (1979). *Antimicrob. Agents Chemother.*, **16**, 240–242.
61. K. Takayama, and J.O. Kilburn, (1984). Proceedings of the US–Japan Leprosy and Tuberculosis Research Conference, Tokyo (in press).
62. K. Takayama, and N. Qureshi, (1984). In *The Mycobacteria: A Sourcebook* (Eds. G.P. Kubica and L.G. Wayne), Part A, pp. 315–344, Marcel Dekker, New York.
63. P.A. Tisdall, G.D. Roberts, and J.P. Anhalt, (1979). *J. Clin. Microbiol.*, **10**, 506–514.
64. A.Y. Tsang, V.L. Barr, J.K. McClatchy, M. Goldberg, I. Drupa, and P.J. Brennan, (1984). *Int. J. Syst. Bacteriol.*, **34**, 35–44.
65. E. Vilkas, C. Amar, J. Markovits, J.F.G. Vliegenthart, and J.P. Kamerling, (1973). *Biochim. Biophys. Acta*, **297**, 423–435.
66. F.G. Winder, (1982). In *The Biology of the Mycobacteria* (Eds. C. Ratledge and J.L. Stanford), pp. 353–438, Academic Press, London.

# 3 The chemotherapy of filarial nematode infections of man: aspirations and problems

D.A. Denham
London School of Hygiene and Tropical Medicine, Keppel Street, London WC1E 7HT, UK
and
J. Barrett
Department of Zoology, University of Wales, Aberystwyth, Dyfed SY23 3DA, Wales

## 3.1 Introduction

Two major types of filarial nematode parasitize man. These are nematode worms of the order Filaroidea which have male and female adults and young forms called microfilariae.

*Wuchereria bancrofti and Brugia malayi* have adults, 5–7 cm long, which live in the lymphatics and microfilariae, 200 μm long, which circulate in the blood. Whilst most infections with these parasites are relatively asymptomatic and produce transient fevers, lymphadenitis (inflamed lymph nodes) and lymphangitis (inflamed lymphatics), sometimes patients develop the grossly disfiguring diseases of elephantiasis and hydrocoele. These diseases are almost certainly immunopathological and the major pathogenic stage of the life cycle is the adult worm in the lymphatics.

*Onchocerca volvulus*, on the other hand, has larger adult worms (the females are 30–50 cm long) which live in subcutaneous nodules and have microfilariae (300 μm long) which live in the skin. Although the nodules may be considered cosmetically unpleasant they pale into insignificance when compared with the dreadful conditions caused by the microfilariae. Once again the disease seems to be immunopathological. The skin lesions give rise to pruritis (which starts mildly but can become devastatingly itchy), depigmentation, loss of elasticity and roughening. More seriously, the microfilariae can invade the eye and cause blindness. In the eye microfilariae can irreversibly damage the cornea, lens, uvea and retina, all of which can lead to a loss of vision.

About 350 million people are infected with lymphatic filariasis and 40 million with onchocerciasis. Lymphatic filariasis is transmitted by mosquitoes and is widely distributed in the wet tropical zones of Africa, Asia, Oceania and America. Onchocerciasis is transmitted by blackflies (*Simulium* species) and occurs in African and Central and South America.

## 3.2 Current chemotherapy

Only one drug has been widely used against the lymphatic filariae. This compound is diethylcarbamazine (hereafter DEC) which is 1-diethylcarbamyl-4-methypiperazine (**1**). It was one of a series of piperazine derivatives developed in response to the acute filariasis suffered by US marines and soldiers during the Second World war whilst they were fighting in the South Pacific. Hewitt *et al.* (1947) published the results of their experiments with different piperazine derivatives and decided

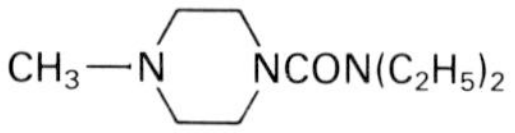

(**1**) Diethyl carbamazine (DEC)

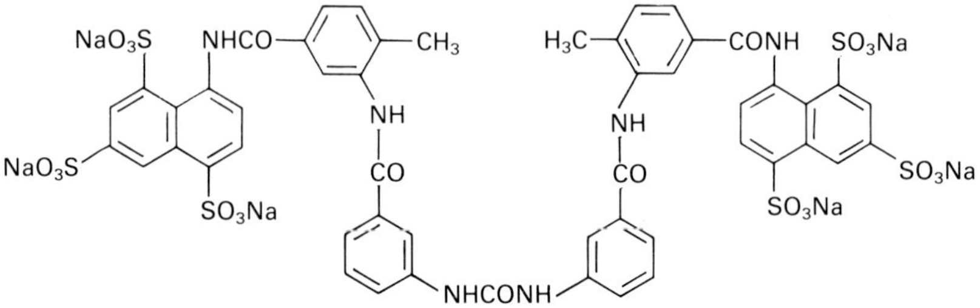

(**2**) Suramin

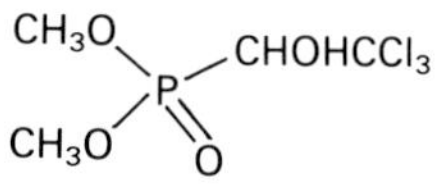

(**3**) Metrifonate

that DEC was the most active of these. Santiago-Stevenson, Oliver-Gonzalez and Hewitt (1947) showed that DEC was highly effective against microfilariae of *W. bancrofti* and Wilson (1950) showed that it had similar effects on *B. malayi*.

It is very simple to examine blood for microfilariae and, therefore, the microfilaricidal effects of DEC are easy to demonstrate. However, it is much more difficult to demonstrate that DEC is macrofilaricidal. This subject has recently been extensively reviewed by Ottesen (1984) who concludes, from all the evidence, that under the right circumstances DEC does kill the adult worms. Thooris (1956) showed that suramin is macrofilaricidal in *W. bancrofti* infected patients but, as will be seen later, its use cannot be justified in this infection.

For onchocerciasis two drugs have been widely used. DEC is unequivocally microfilaricidal, as was demonstrated originally by Mazzoti (1948). There is no evidence that DEC kills adult *O. volvulus*. The other drug widely used for the chemotherapy of onchocerciasis, suramin (**2**), is definitely macrofilaricidal (Duke, 1968). The effects of suramin were first reported by van Hoof and coworkers (1947) who noticed that after treating patients who had onchcerciasis, skin microfilarial levels were low about a year after treatment. This was because the adult worms had been killed and no new microfilariae had been born to replace those dying of old age.

The organophosphate, metrifonate (**3**), has also been used for the treatment of onchocerciasis (Salazar-Mallen, Gonzalez-Barranco and del Carmen Moutes, 1971). It is microfilaricidal but as it kills fewer microfilariae than does DEC the Mazzoti reactions (see below) produced are less severe.

### 3.3 Side effects of treatment

The previous section may suggest that effective chemotherapy exists for the filariae. However, two features were deliberately ignored: treatment schedules and side effects.

DEC has to be given on several occasions with a total dosage of 50–72 mg/kg. The ways in which this is carried out are legion. Side effects of DEC treatment are often severe. Wilson (1950) shows that the body temperatures of patients with *B. malayi* who were treated with DEC rose dramatically during treatment. Reactions in *B. timori* infections are even more severe (Partono, Purnomo and Soewarta, 1979) and were most graphically illustrated in the following words: 'There was no laughter or loud voices of playing children; the whole village was quiet and an eerie feeling came over the village as if it had been abandoned by the population'.

Painful swellings have also been reported in the lymphatics, especially those in the scrotum (Santiago-Stevenson, Oliver-Gonzalez and Hewitt, 1947; Hawking and Smith, 1952; Mak and Zaman, 1980). These are probably associated with the death of the adult worm (Ottesen, 1984).

The side effects of DEC in lymphatic filariasis seem almost trivial when compared with the aftermath of the chemotherapy of onchocerciasis. DEC kills the microfilariae in the skin and the subsequent release of antigen produces an intense pruritis. This is extremely unpleasant and so specific that it is sometimes used as a diagnostic technique, when it is called the Mazzoti reaction after its discoverer (Mazzoti, 1948). There have even been reports of death associated with DEC treatment of some severely affected patients.

The side effects of suramin are even worse and numerous deaths have been associated with its use. Unfortunately the cause(s) of death has (have) been difficult to establish and deaths occurred at varying times after treatment. It is therefore not possible to predict, with any certainty, the toxic effects of therapy.

Because of the paucity of compounds which can be used to treat onchocerciasis a great deal of research has been, and is being, carried out on dosage regimens for suramin. However, the most striking features of the chemotherapy with suramin are its problems: it has to be injected intravenously, it should only be given under direct qualified medical supervision, numerous injections are required and close observation, including laboratory analysis of specimens from the patient, is needed during treatment.

The organic arsenical, Mel-W, has also been used to treat onchocerciasis but it appears to have even more unpredictable side effects than suramin (Duke, 1970). A small number of fatalities have occurred with Mel-W. This has inhibited the further development of other arsenic compounds.

### 3.4 What is needed?

DEC is cheap and relatively effective against *Brugia* and *Wuchereria* and has been successfully used virtually to eliminate *Wuchereria* from large areas of the South Pacific (Hawking and Denham, 1976). However, it has to be administered frequently and its initial side effects mitigate against full patient compliance. This can be overcome by careful education of the infected populace, but is still a major obstacle.

One form of *B. malayi* is a zoonosis with monkeys acting as a reservoir host. DEC could be used as a chemoprophylactic in the area where this occurs. Again frequent use of DEC would be necessary and an even better result could be obtained with a depot drug which killed all developing larvae.

There can be no doubt that a new chemotherapy is needed for *Onchocerca* infections. The question which arises is: 'What should be the target stage of the life cycle?' The WHO have recommended that top priority should be given to the development of an antiadult (macrofilaricidal) compound: to this, we would add that, if possible, this compound should kill all developing stages so that it could be used as a chemoprophylactic. A major problem of chemotherapy is the side effects which arise from allergic reactions to the filarial enzymes and other antigenic material liberated when dead adult filariae are broken down in the body. Many patients are infected with large numbers of adult worms and therefore suffer from extensive side effects during therapy. The separation of these side effects from those directly attributable to the drug is virtually impossible.

The same problems are also associated with the killing of microfilariae, the major pathogen, which give rise to the Mazzoti reaction. The newly introduced microfilaricide, ivermectin (**4**), is interesting in that reported side effects of the Mazzoti type are minimal (Aziz *et al.*, 1982; Awadzi *et al.*, 1984).

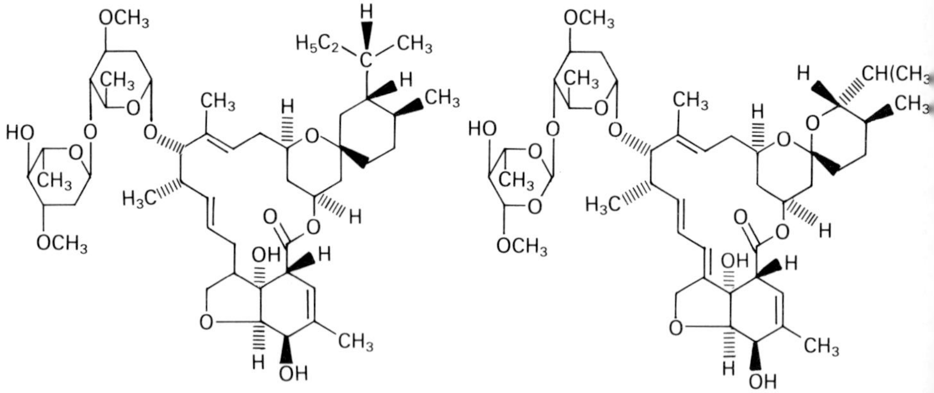

Avermectin $B_1a$ (80%) Avermectin $B_{1b}$ (20%)

(**4**) Ivermectin

### 3.5 Biochemistry of filarial nematodes

Compared with many other parasitic nematodes, relatively little is known about the biochemistry of the Filaroidea. Much of the work on filarial worms has involved species of non-medical importance and the extrapolation of results from one species of filariid to another may not be justified.

### 3.5.1 *Catabolic reactions*

*(a) Glycolysis* As in other parasitic nematodes carbohydrate is the major and probably the sole energy source for adult filarial worms. Unlike most helminths, filaria have a relatively low glycogen content — usually less than 2 per cent of their fresh weight (Barrett, 1983).

Adult filarial worms have a normal glycolytic sequence, and glycolytic enzymes have been demonstrated in whole or in part in *B. pahangi*, *B. malayi*, *Chandlerella hawkingi*, *Dipetalonema vitae*, *Dirofilaria immitis*, *Litosomoides carinii*, *O. volvulus* and *Setaria cervi* (Barrett, 1983; Flockhart and Denham, 1984; Oothuman, Moss and Maddison, 1984).

There have been no systematic studies on the regulatory enzymes of glycolysis in filarial worms. Filariid hexokinases appear to be relatively non-specific with regards to their substrate, the enzyme from adult *D. immitis* phosphorylating glucose, fructose, mannose and glucosamine (Hutchison, Turner and Oelshlegel, 1977), that from adult *L. carinii* and *C. hawkingi* glucose, fructose, mannose and galactose (Srivastava and Ghatak, 1971). *In vitro* both *S. cervi* and *C. hawkingi* preferentially utilize exogenous glucose or mannose for maintenance rather than fructose or galactose (Srivastava, Ghatak and Krishna Murti, 1968; Srivastava and Ghatak, 1974; Anwar, Srivastava and Ghatak, 1978). The hexokinase of *D. immitis* is composed of at least three isoenzymes and is only weakly inhibited by glucose-6-phosphate (Hutchison *et al.*, 1978). The pyruvate kinases of adult *L. carinii* and *D. immitis* have a low affinity for phosphoenolpyruvate, are activated by fructose-1,6-diphosphate and are inhibited by ATP; in this they resemble the vertebrate L-type enzyme (Brazier and Jaffe, 1973). The pyruvate kinase from adult *D. immitis* has a molecular weight of 270 000 and requires a divalent cation for activity; the enzyme is also activated by $K^+$ or $NH_4^+$ and ADP can be replaced by GDP, IDP, UDP, CDP or *d*-ADP (in order of decreasing efficiency). The pyruvate kinase of *O. volvulus* is also activated by fructose-1,6-diphosphate and inhibited by ATP and malate (Walter and Van den Bossche, 1980). Nothing is known about the control of the other regulatory enzymes of glycolysis in filariids, namely phosphorylase and phosphofructokinase. The phosphorylase of *L. carinii* exists in **a** and **b** forms (Nelson and Saz, 1982), whilst the phosphofructokinase of adult *L. carinii*, *D. vitae* and *B. pahangi* are considerably more sensitive to inhibition by trivalent antimonial compounds than is the corresponding mammalian enzyme. Protein kinases I, II and III have been demonstrated in *O. volvulus* (Walter and Schulz-Key, 1980a–c), but their role in metabolic regulation in filariids has yet to be established.

Adult *D. vitae* and *B. pahangi* are homolactic fermentors, under both aerobic and anaerobic conditions, i.e. they break down carbohydrate to lactate only (Wang and Saz, 1974; Saz and Dunbar, 1975); *C. hawkingi* is similar and during aerobic incubations 85–100 per cent of the carbohydrate catabolized by this parasite can be accounted for as lactate, with traces of pyruvate (Srivastava, Ghatak and Krishna Murti, 1968). In *D. immitis*, lactate accounts for some 55 per cent of the glucose metabolized during short-term incubations (1 h), which increases to 93 per cent during long-term incubations (24 h); only traces of acetate and no propionate or acetoin were detected (McNeill and Hutchison, 1972; Hutchison and Turner, 1979a; Turner and Hutchison, 1983). Adult *D. uniformis* is also a homolactic fermentor (von Brand *et al.*, 1963).

The nature of the end products produced by adult *L. carinii*, however, varies with the oxygen tension. *Litosomoides carinii*, unlike *D. viteae* and *B. pahangi*, is an obligate aerobe (Wang and Saz, 1974). Under aerobic conditions 30–40 per cent of the carbohydrate catabolized by adult *L. carinii* can be accounted for as lactate and 25–35 per cent as acetate, and there are small amounts of acetoin and traces of pyruvate (Bueding, 1949a; Berl and Bueding, 1951). Anaerobically, 80 per cent of the carbohydrate is metabolized by adult *L. carinii* to lactate, the remainder to acetate (Bueding, 1949a). The latter is formed by the decarboxylation of pyruvate and this almost certainly involves the pyruvate dehydrogenase complex, which has a high activity in *L. carinii* but only a low activity in *B. pahangi* and *D. viteae* (Wang and Saz, 1974; Middleton and Saz, 1979). Aerobically, *L. carinii* excretes three times as much acetate as anaerobically. The pyruvate dehydrogenase complex yields acetylCoA, which is then cleaved to acetate; this route of acetate formation is widespread in helminths and may be coupled to ATP production (Barrett, 1981, 1984). A number of different pathways have been suggested for acetoin production in *L. carinii* (Berl and Bueding, 1951); the most likely source, however, is from a partial reaction of the pyruvate dehydrogenase complex (Middleton and Saz, 1979; Barrett and Butterworth, 1984). Adult *L. carinii* can utilize exogenous acetoin under aerobic and much more slowly under anaerobic conditions (Berl and Bueding, 1951).

Adult *S. cervi* show a pattern of excretory products similar to that of *L. carinii*; aerobically only 28 per cent of the metabolized glucose can be accounted for as lactate and the parasite produced significant amounts of acetoin (Yonezawa, 1952, 1953; Pandya, 1961; Anwar, Ansari and Ghatak, 1975).

*(b) Carbon dioxide fixation* Phosphoenolpyruvate carboxykinase and the malic enzyme have both been reported from adult filarial worms, but pyruvate carboxylase appears to be absent. Phosphoenolpyruvate carboxykinase has been demonstrated in *C. hawkingi*, *D. immitis*, *L. carinii* and *S. cervi*, whilst the malic enzyme has been reported from *B. pahangi*, *C. hawkingi*, *D. immitis* and *O. volvulus* (Barrett, 1983; Turner and Hutchison, 1983).

The ratio of the activities of pyruvate kinase to phosphoenolpyruvate carboxykinase is often taken by helminth biochemists as an indicator of the relative importance of lactate production on the one hand and of the partial reversed tricarboxylic acid cycle of the type found in *Ascaris* (leading to succinate and propionate production) on the other (Figure 3.1). High ratios (greater than 5)

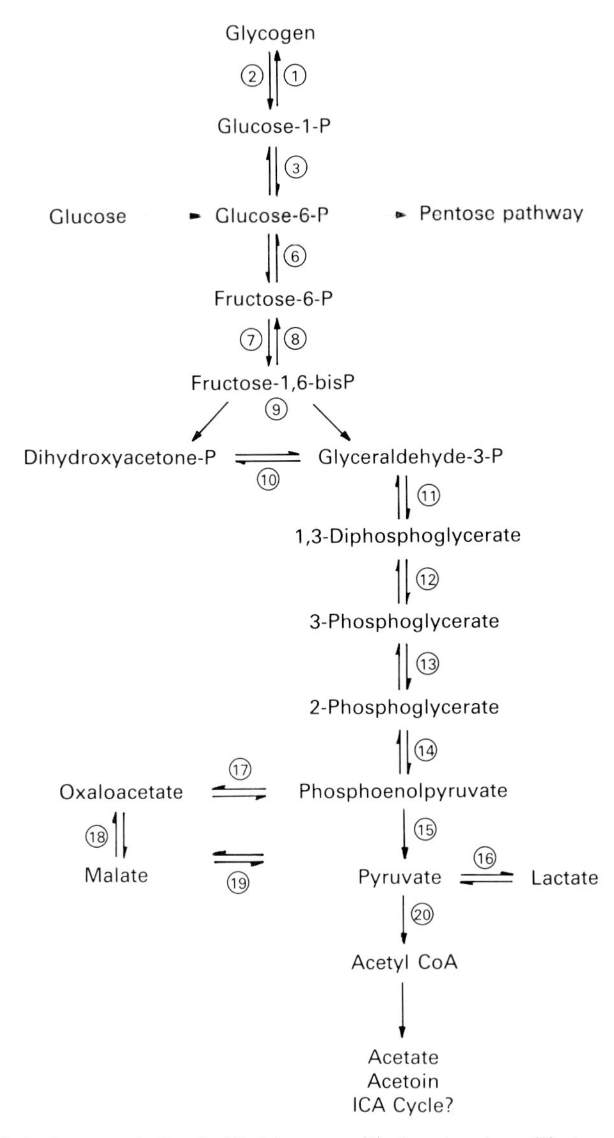


**Figure 3.1** Carbohydrate catabolism in filarial worms: (1) phosphorylase, (2) glycogen synthase, (3) phosphoglucomutase, (4) hexokinase, (5) glucose-6-phosphate dehydrogenase, (6) glucose-phosphate isomerase, (7) phosphofructokinase, (8) fructose-1,6-bisphosphatase, (9) aldolase, (10) triosephosphate isomerase, (11) glyceraldehyde-3-phosphate dehydrogenase, (12) phosphoglycerate kinase, (13) phosphoglyceromutase, (14) enolase, (15) pyruvate kinase, (16) lactate dehydrogenase, (17) phosphoenolpyruvate carboxykinase, (18) malate dehydrogenase, (19) malic enzyme, (20) pyruvate dehydrogenase.

are characteristic of homolactic fermentors whilst low ratios (less than 1) are found in helminths that produce propionate or succinate. In adult filarial worms the ratio varies from 10 in *D. immitis* (Brazier and Jaffe, 1973), 7.8 in *C. hawkingi* (Srivastava and Ghatak, 1971), 2.8–0.5 in *L. carinii* (Brazier and Jaffe, 1973, Srivastava *et al.*, 1970) to 0.4 in *S. cervi* (Anwar *et al.*, 1977a).

In species with high phosphoenolpyruvate carboxykinase activities, malate formed by the carboxylation of phosphoenolpyruvate could be cleaved by the malic enzyme to pyruvate. This would result in separate cytoplasmic and mitochondrial pyruvate pools, leading to lactate and acetate formation respectively. Separate pyruvate pools for lactate and acetate production would greatly simplify metabolic control. However, apart from an unconfirmed report of fumarate reductase activity in *S. cervi* (Anwar *et al.*, 1977a) there is no evidence that carbon dioxide fixation and a partial reverse tricarboxylic acid cycle of the type found in *Ascaris* is important in filarial worms. Both phosphoenolpyruvate carboxykinase and the malic enzyme have important synthetic functions. Phosphoenolpyruvate carboxykinase is involved in the resynthesis of glycogen from pyruvate (itself derived from lactate or alanine). The high levels of this enzyme in certain filarial worms may be related not to phosphoenolpyruvate carboxykinase's catabolic role, but to its synthetic function in neoglycogenesis, coupled with the low glycogen content of these parasites. The malic enzyme generates NADPH for synthetic reactions and may also be involved in a cycle with citrate lyase and citrate synthase transporting acetylCoA across the mitochondrial membrane (Stryer, 1981). This may be particularly important in parasites which excrete acetate. The malic enzyme could also be the source of labelled carbon dioxide produced during incubation with $^{14}$C-glucose in parasites which lack a functional tricarboxylic acid cycle.

*(c) Tricarboxylic acid cycle* A more or less complete sequence of tricarboxylic acid cycle enzymes has been demonstrated in adult *B. pahangi* and *D. immitis*, whilst individual enzymes have been reported from *C. hawkingi*, *D. viteae*, *L. carinii*, *O. volvulus*, *S. cervi* and *S. digitata* (Barrett, 1983; Khatoon, Wajihullah and Ansari, 1983). The activities of two of the enzymes at the beginning of the cycle, aconitase and isocitrate dehydrogenase, were extremely low in *D. immitis* and *L. carinii* (McNeill and Hutchison, 1971; Ramp and Köhler, 1982) and the latter enzyme could not be detected at all in *C. hawkingi* (Srivastava, Ghatak and Krishna Murti, 1968). Wang and Saz (1974) concluded, on the basis of isotope studies, that the tricarboxylic acid cycle was not a significant pathway for carbohydrate degradation in adult *L. carinii*, *B. pahangi* or *D. viteae* and the same seems to be true for *D. immitis* and *O. volvulus* (McNeill and Hutchison, 1971; Walter and Vanden Bossche, 1980; Ramp and Köhler, 1982).

Fluoroacetate, a classical inhibitor of the tricarboxylic acid cycle, causes a decrease in oxygen uptake and motility in *L. carinii* (Bueding, 1949a). This cannot, however, be taken as definitive evidence for a functional tricarboxylic acid cycle in this parasite; fluoroacetate derivatives can inhibit a variety of enzymes including those concerned with the conversion of pyruvate to acetate.

*(d) Pentose phosphate pathway* There is both isotopic and enzymatic (glucose-6-phosphate dehydrogenase, 6-phosphogluconate dehydrogenase, *trans*-aldolase, *trans*-ketolase) evidence for a pentose phosphate pathway in *D. immitis* (Hutchison and McNeill, 1979; Turner and Hutchison, 1983). Glucose-6-phosphate dehydrogenase and 6-phosphogluconate dehydrogenase have also been shown in adult *S. cervi* (Anwar *et al.*, 1977a, b); however, neither of these two enzymes could be detected in adult *C. hawkingi* (Srivastava, Ghatak and Krishna Murti, 1968).

*(e) Lipid and amino acid catabolism* Nothing is known about lipid breakdown in filarial worms, the only report being for adult *B. pahangi* which, in common with other parasitic nematodes, is unable to oxidize exogenous palmitic acid (Middleton and Saz, 1979). Even less is known about amino acid catabolism in filarial worms. Govindwar, Gawande and Harinath (1974) failed to detect amino transferase activity in *W. bancrofti*, but glutamate dehydrogenase, an enzyme which plays a central role in amino acid deamination (as well as being the major pathway for the formation of amino groups from ammonia), has been demonstrated in the microfilariae of *D. immitis* (Langer and Jiampermpoon, 1970).

*(f) Larval stages* Compared to the adult worms, little is known of the catabolic reactions of the larval stages of filariids and what work there is has been done with microfilariae, nothing being known about the stages in the arthropod vector. Most of the enzymes of glycolysis and the tricarboxylic acid cycle have been demonstrated in the microfilariae of *S. cervi*, the pyruvate kinase/phosphoenolpyruvate carboxykinase ratio being 0.37 (Rathaur *et al.*, 1982). Pyruvate kinase and lactate dehydrogenase have been demonstrated in the microfilariae of *D. immitis* and the regulatory properties of the pyruvate kinase are said to be similar to those of the adult (Langer and Jiampermpoon, 1970; Brazier and Jaffe, 1973). Under anaerobic conditions, the microfilariae of *B. pahangi* convert 99 per cent of labelled glucose to lactate; in air, 80 per cent goes to lactate, the rest to acetate (presumably via the pyruvate dehydrogenase complex), with a low level of complete oxidation, the latter possibly involving a tricarboxylic acid cycle or the pentose pathway and malic enzyme (Rew and Saz, 1977). In the microfilariae of *D. immitis* some 40 per cent of labelled glucose finishes up as carbon dioxide under aerobic conditions.

There is some isotopic and enzymic evidence for a pentose phosphate pathway in the microfilariae of *D. immitis* (Jaffe and Doremus, 1970; Langer and Jiampermpoon, 1970) and both glucose-6-phosphate dehydrogenase and 6-phosphogluconate dehydrogenase have been found in the microfilariae of *S. cervi* (Rathaur *et al.*, 1982).

### *3.5.2 Terminal oxidation*

*(a) Respiratory studies* Like other parasitic helminths, filarial worms all utilize oxygen when it is available. Oxygen uptake has been measured in the microfilariae and adults of *D. immitis* and in adult *D. uniformis* and *L. carinii* (Bueding, 1949a,

1949b, 1949c; von Brand, 1960; von Brand *et al.*, 1963; Jaffe and Doremus, 1970). The microfilariae of *D. immitis* show no Pasteur effect and respiration is stimulated by exogenous glucose. The microfilariae of *B. pahangi* show a reverse Pasteur effect, i.e. less glucose is catabolized under anaerobic than aerobic conditions (Rew and Saz, 1977). Adult *D. immitis* show only a very small Pasteur effect, whilst adult *L. carinii* show a marked Pasteur effect, but no oxygen debt (Bueding, 1949a, 1949b, 1949c; Hutchison and Turner, 1979a, 1979b). Oxygen uptake in *L. carinii* is stimulated by exogenous glucose, fructose and mannose, but not galactose. Respiration in adult *L. carinii* is oxygen dependent, but relatively unaffected by pH or ionic balance (Bueding, 1949a, 1949b). In the absence of glucose the RQ of adult *L. carinii* was 0.44; in the presence of glucose, 0.94.

Filarial worms differ from one another in their oxygen requirements. Adult *L. carinii* require oxygen for survival and motility, and they are obligate aerobes; on the other hand, adult *D. viteae* and adult *B. pahangi* can remain motile under nitrogen for at least 24 hours (Wang and Saz, 1974). Similarly adult *D. immitis* will remain alive and active for at least 24 hours under anaerobic conditions (Hutchison and Turner, 1979a), although Earle (1959) found that *D. immitis* required oxygen for long-term survival during *in vitro* culture. Rew and Saz (1977) found that the microfilariae of *L. carinii*, *D. viteae* and *B. pahangi* all required oxygen for movement, but not necessarily for survival.

*(b) Nature of terminal oxidase* Normal-looking cristate mitochondria have been described in adult *B. pahangi* and *L. carinii* and in the adults and microfilariae of *D. immitis* (Lee and Miller, 1967, 1969; Johnson and Bemrick, 1969; Middleton and Saz, 1979). Early work on the terminal oxidase of filarial worms suggested that cytochromes were absent and that the oxidase was possibly a flavoprotein. Thus, neither cytochrome **c**, nor cytochrome **c** oxidase was detected in adult *L. carinii* (Bueding and Charms, 1951, 1952; Wang and Saz, 1974) and nor could cytochrome oxidase be detected in adult *B. pahangi* or *D. viteae* (Wang and Saz, 1974). Oxygen uptake in *L. carinii* is inhibited by cyanide and cyanine dyes, and in both cases this is accompanied by a compensatory increase in glycolysis and lactate production (Welch *et al.*, 1947; Bueding, 1949a, 1949b; Peters *et al.*, 1949). A variety of sites of action have been described for cyanine dyes and the drug may not necessarily be inhibiting the terminal oxidase. Inhibition of the pyruvate dehydrogenase complex, for example, would lead to a decrease in acetate formation and decreased oxygen uptake, with a compensatory increase in lactate production.

More recently cytochrome oxidase activity has been demonstrated in the microfilariae of *B. pahangi* (Rew and Saz, 1977), and Hayashi and Oya (1978) have reported the presence of cytochrome **c**, two **b**-type cytochromes and low levels of cytochrome **a** in adult *D. immitis*. These authors also suggested that adult *D. immitis* may have a branched cytochrome chain, with an **o**-type cytochrome as the alternative oxidase. On the basis of inhibitor studies Ramp and Köhler

(1982) have deduced that *L. carinii* does possess a functional cytochrome chain and is capable of oxidative phosphorylation; a similar conclusion has been reached for adult *B. pahangi* and *D. viteae* (Mendis, unpublished data). Oxidative phosphorylation has also recently been demonstrated in body wall preparations from *S. cervi* (Singh, Pampori and Srivastava, 1984).

There is no doubt that filarial worms possess functional cytochrome chains. What is not known is the relative contribution of oxidative processes to the overall energy balance of the parasite, and this may well vary from species to species and at different stages of development.

### 3.5.3 *Synthetic reactions*

In general synthetic pathways are much more poorly known in parasitic helminths than catabolic pathways. The information tends to be patchy; some pathways are known in great detail whilst other areas have been almost totally ignored.

*(a) Folate metabolism* As a potential target site, folate metabolism has proved very successful in the chemotherapy of microorganisms; the folate metabolism of filarial worms has been investigated in great detail by Jaffe and his coworkers.

The majority of enzymes involved in the interconversion of folate analogues have been demonstrated in filariids (Figure 3.2). Adult *B. pahangi* and *D. immitis* have been shown to have $N^5$,$N^{10}$-methylenetetrahydrofolate reductase, serine hydroxymethyltransferase, $N^5$,$N^{10}$-methylenetetrahydrofolate dehydrogenase, $N^5$,$N^{10}$-formyltetrahydrofolate synthetase, NADP-dependent and NADP-

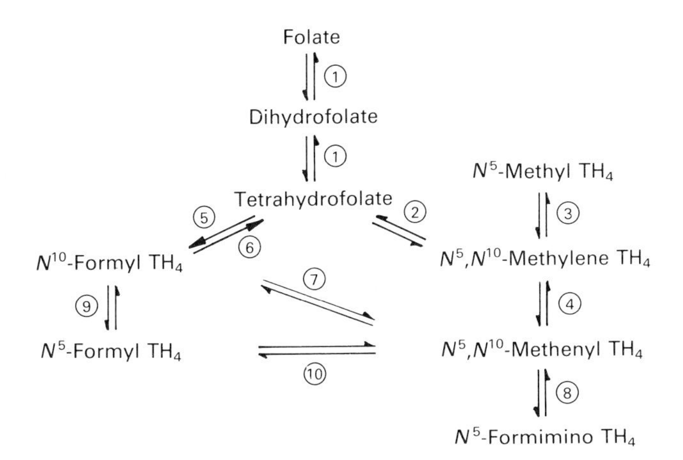


**Figure 3.2** Folate metabolism: (1) dihydrofolate reductase, (2) serine hydroxymethyltransferase, (3) $N^5$,$N^{10}$-methylenetetrahydrofolate reductase, (4) $N^5$,$N^{10}$-methylenetetrahydrofolate dehydrogenase, (5) $N^{10}$-formyltetrahydrofolate synthetase, (6) $N^{10}$-formyltetrahydrofolate dehydrogenase, (7) $N^5$,$N^{10}$-methenyltetrahydrofolate cyclohydrolase, (8) $N^5$-formiminotetrahydrofolate cyclodeaminase, (9) $N^5$-formyl,$N^{10}$-formyltetrahydrofolate mutase, (10) $N^5$-formyltetrahydrofolate cyclodehydrase.

independent $N^{10}$-formyltetrahydrofolate dehydrogenase, $N^5,N^{10}$-methenyltetrahydrofolate cyclohydrolase, $N^5$-formyltetrahydrofolate cyclodehydrolase, $N^5$-formiminotetrahydrofolate cyclodeaminase and formiminoglutamate:tetrahydrofolate $N^5$-formiminotransferase; there is also indirect evidence for $N^5$-formyl, $N^{10}$-formyltetrahydrofolate mutase (Jaffe and Chrin, 1980, 1981; Jaffe, Chrin and Smith, 1980; Comley *et al.*, 1981). The cofactor requirements of $N^5,N^{10}$-methylenetetrahydrofolate reductase have been studied in some detail and this enzyme (which is a flavoprotein) is unusual in filarial worms in that it works preferentially in the reverse direction to that in mammals, i.e. in the direction of $N^5,N^{10}$-methylenetetrahydrofolate formation (Jaffe, 1980a; Comley and Jaffe, 1981; Comley *et al.*, 1981). Mammalian cells cannot convert $N^5$-methyltetrahydrofolate back to $N^5,N^{10}$-methylenetetrahydrofolate and the $N^5$-methyl analogue is the major folate derivative found in mammalian tissue. Possibly the situation in filariids is an adaptation by the parasites to make use of this host folate source.

Dihydrofolate reductase has been shown in several adult filarial worms (*D. immitis*, *L. carinii*, *D. viteae* and *O. volvulus*) and the sensitivity of the enzyme to possible inhibitors has been investigated (Jaffe, 1971, 1972; Jaffe, McCormack and Meymarian, 1972). There is good indirect evidence that adult filarial worms cannot synthesize dihydrofolate *de novo* and require a source of preformed folates (Jaffe, 1980a). There are, however, conflicting reports as to the ability of filarial worms to take up folate derivatives. Jaffe (1980a) found that adult *B. pahangi* could assimilate and anabolize exogenous folic acid and to a lesser extent $N^5$-formyltetrahydrofolate; however, during short-term incubations Chen and Howells (1981a) could detect no uptake of folic acid by the adults, juveniles, infective larvae or microfilariae of *B. pahangi*.

In contrast to the adult worms, no dihydrofolate reductase activity was found in the microfilariae of *B. pahangi* or *D. immitis* (Jaffe, 1972; Jaffe, McCormack and Meymarian, 1972; Jaffe *et al.*, 1977). Mosquitoes (*Aedes aegypti*) infected with the larval stages of *B. pahangi* show characteristic changes in the levels of their folate metabolizing enzymes, including an increase in the activities of methionine synthetase, dihydrofolate reductase, serine *trans*-hydroxymethylase and $N^5,N^{10}$-methylenetetrahydrofolate reductase and a decrease in the activity of $N^{10}$-formyltetrahydrofolate synthetase. $N^5,N^{10}$-Methylenetetrahydrofolate dehydrogenase, thymidylate synthetase, xanthine dehydrogenase and lactate dehydrogenase remain unchanged (Jaffe *et al.*, 1977; Jaffe and Chrin, 1978a, 1978b, 1979a, 1979b, 1979c). These results suggest that the developing larvae may be utilizing host folate derivatives (probably $N^5$-methyltetrahydrofolate) and this results in a compensatory increase in the activities of host enzymes involved in $N^5$-methyltetrahydrofolate synthesis.

*(b) Amino acids and proteins* The incorporation of amino acids into proteins has been demonstrated in a number of filarial worms, but nothing is known about the details of protein synthesis. Serine hydroxymethyltransferase, the enzyme

responsible for the interconversion of serine and glycine, has been demonstrated in adult *B. pahangi* and *D. immitis* (Jaffe and Chrin, 1980). However, neither methionine synthase nor betaine:homocysteine *trans*-methylase could be demonstrated in adult *B. pahangi* and both *B. pahangi* and *D. immitis* probably require an exogenous source of methionine (Jaffe, 1980b; Jaffe and Chrin, 1979a, 1981). These two parasites can convert methionine to cysteine, using the pathway via 5-adenosylmethionine, 5-adenosylhomocysteine, homocysteine and cystathione (Jaffe, 1980b).

Amino acid metabolism of helminths is a very neglected area and there is still much to be learnt about the amino acid requirements of parasites.

*(c) Nucleotides and nucleic acids* Studies suggest that filarial worms have both *de novo* and salvage pathways for purine and pyrimidine biosynthesis. *In vitro* the uptake of radioactively labelled adenine, adenosine, uracil, uridine, hypoxanthine and guanine has been demonstrated in a variety of filariae, but in no case was there any uptake of thymine, thymidine or cytosine (see transport mechanisms). The actual incorporation of uridine and uracil into nucleic acids has only been reported in adult *D. immitis* (Jaffe, McCormack and Meymarian, 1972) and in the microfilariae which incorporated uridine, uracil, adenine and adenosine into RNA (Jaffe and Doremus, 1970). The inability to take up preformed cytosine, thymine or thymidine *in vivo* would suggest that salvage pathways for these pyrimidines might be lacking in filarial worms. However, more recently Jaffe, Comley and Chrin (1982) have shown incorporation of thymidine into DNA *in vitro* in *B. pahangi* and have demonstrated the initial enzyme of the salvage pathway, thymidine kinase, in extracts of adult female *B. pahangi* and *D. immitis*.

There are a number of reports supporting *de novo* synthesis of pyrimidines and purines in filarial worms. Orotate derivatives are incorporated into RNA by the microfilariae of *D. immitis*, although there was no incorporation of glycine or formate into nucleic acids (Jaffe and Doremus, 1970). Incubation of adult *B. pahangi* with labelled methyltetrahydrofolate resulted in labelled adenosine and guanine ribonucleotides as well as labelled inosine monophosphate (Jaffe and Chrin, 1981). The label was evenly distributed among the different nucleotides, showing that *B. pahangi* adults interconvert purine nucleotides with IMP as the common intermediate. Evidence for *de novo* pyrimidine synthesis in filarial worms also comes from the demonstration of thymidylate synthetase in adult *D. immitis* and *B. pahangi* (Jaffe and Chrin, 1981). Finally, the incorporation of label from glycine into DNA and RNA (and of inorganic phosphate into RNA) has been reported in adult *L. carinii* (Akinwande and Akinsimisi, 1980).

*(d) Lipids* All of the major neutral and phospholipid fractions have been demonstrated in filarial worms (Barrett, 1983). Unusually large amounts of plasmalogens and phosphatidylalkyl ethers have been reported to occur in adult *L. carinii* (Subrahmanyam, 1967), but the significance of this observation is obscure.

In common with other parasitic helminths, filarial worms can synthesize all of their complex lipids from simple precursors. The microfilariae of *D. immitis* incorporated label from radioactive glucose into their total lipid fraction (Jaffe and Doremus, 1970), whilst labelled glycerol, acetate and oleate were incorporated into the total lipid, phospholipid and diacylglycerol fractions of adult *D. immitis* (Hutchison and Turner, 1979b; Turner and Hutchison, 1979). Adult *D. immitis* and *B. pahangi* incorporated label from $^{14}$C-mevalonate into triacylglycerols, free fatty acids, sterol esters, phosphatidylethanolamine, phosphatidylcholine, lysophosphatidylcholine, phosphatidylinositol, phosphatidylserine and sphingomyelin (Comley and Jaffe, 1981; Comley *et al.*, 1981). Although, like other parasitic nematodes, filarial worms are unable to synthesize long chain fatty acids *de novo*, the incorporation of label from mevalonate into free fatty acids suggests that these filarial worms have a fatty acid lengthening system, presumably involving acetylCoA.

As well as being unable to synthesize long chain fatty acids *de novo* filarial worms, in common with other parasitic nematodes, are also unable to synthesize squalene or sterols *de novo*. However, adult *B. pahangi* and *D. immitis* can synthesize ubiquinone 9 and short and long chain isoprenoid alcohols from mevalonate (Comley and Jaffe, 1981; Comley, *et al.*, 1981). Although these worms could synthesize ubiquinone, neither of the two parasites could convert menadione (vitamin $K_3$) to menaquinone (vitamin $K_2$). The short chain isoprenoid alcohol fraction synthesized by *B. pahangi* and *D. immitis* consisted mainly of geranyl geraniol, whilst the long chain isoprenoid alcohol fraction was a family of dolichol isoprenologs, including dolichol 18 ($C_{90}$), dolichol 19 ($C_{95}$), dolichol 20 ($C_{100}$), dolichol 21 ($C_{105}$) and dolichol 22 ($C_{110}$). In contrast, farnesol ($C_{15}$) could not be detected, but a band corresponding to the $C_{55}$ isoprenolog was labelled.

Dolichols and related isoprenols act as intermediate lipid carriers in the enzymatic transfer of sugar groups to glycoproteins and proteoglycans; a microsomal glycosyltransferase system has been identified in both *D. immitis* and *B. pahangi* (Comley and Jaffe, 1981; Comley *et al.*, 1981; Comley, Jaffe and Chrin, 1982). High levels of retinol have recently been reported in *O. volvulus* (Sturchler, Wyss and Hank, 1981) and Comley and Jaffe (1983) have shown that *B. pahangi* rapidly takes up retinol and β-carotene. The β-carotene is metabolized to retinol and the latter metabolized to a variety of retinoids tentatively identified as retinylphosphate, retinylphosphate mannose and anhydroretinol. Retinylphosphate may also function as an intermediate lipid carrier in filarial glycoprotein synthesis.

*(e) Complex carbohydrates* Filarial worms can synthesize glycogen from hexose precursors; the role of non-hexose precursors has not been investigated. Glucose and mannose, but not galactose, are glyconeogenic substrates for adult *L. carinii*, but there are conflicting reports as to the role of fructose (Bueding, 1949a; Srivastava, Ghatak and Krishna Murti, 1970; Anwar *et al.*, 1977b). Glucose-6-

phosphate-dependent and glucose-6-phosphate-independent forms of glycogen synthase (D and I forms) have been demonstrated in adult *L. carinii* (Komuniecki and Saz, 1982). In keeping with the aerobic character of *L. carinii*, glycogen synthesis occurs only under aerobic conditions. The incorporation of glucose into glycogen has also been demonstrated in the microfilariae of *D. immitis*, *S. cervi* and *L. carinii* (Jaffe and Doremus, 1970; Rathaur *et al.*, 1980).

Glycoproteins have been identified on the cuticles of the microfilariae of *D. immitis* and *B. pahangi* (Cherian *et al.*, 1980; Furman and Ash, 1983), the cuticle of *O. volvulus* adults (Deas, Aguilar and Miller, 1974) and in the intestine of adult *D. immitis* (Lee and Miller, 1969). A glycosyltransferase system involved in glycoprotein synthesis has been found in adult *B. pahangi* and *D. immitis* (see lipid synthesis). The dynamic nature of the surface of the nematode cuticle means that these surface glycoproteins may have an important role in immune responses, and intensive efforts are being made to characterize the surface cuticular components from nematodes (Philipp and Rumjaneck, 1984).

### *3.5.4 Transport mechanisms*

*(a) Role of the intestine* With the exception of the microfilariae and the third-stage infective larvae, filariids have a functional gut. However, although oral feeding has been demonstrated *in vivo* (Howells and Chen, 1981) no one has yet succeeded in demonstrating it *in vitro* (Howells, 1980; Chen and Howells, 1979a, 1979b, 1981b). The absence of oral feeding *in vitro* points to the requirement for a specific feeding stimulus, and this has implications in relation to the route of uptake of anthelmintics during *in vitro* incubations.

*(b) Role of body wall* The adults and larval stages of filarial worms can all, at least *in vitro*, take up low molecular weight substances across the cuticle, which together with the hypodermis shows structural modifications (see Howells, 1980). Cuticular transport is not a usual feature of animal parasitic nematodes.

Adult *B. pahangi* has been shown to be able to take up D-glucose, L-leucine, glycine, cycloleucine and adenosine across the cuticle (Chen and Howells, 1979a, 1979c; Nduka and Howells, 1980; Howells and Chen, 1981), whilst *D. immitis* adults have been shown to take up D-glucose and adenosine by this route (Yanagisawa and Koyama, 1970; Chen and Howells, 1981a). The transcuticular uptake of glycine has also been demonstrated in *Onchocerca guttuerosa* (Howells, 1980). Cuticular transport in *D. immitis* and *B. pahangi* appears to be selective; L-glucose, sucrose or thymidine, for example, are not taken up, but nothing at all is known about the nature of the transport mechanism. The kinetics of the transcuticular uptake of glycine and cycloleucine by *B. pahangi* suggests simple diffusion, but a mediated system present in the hypodermis could well be masked if diffusion of substrate through the cuticle was the rate-limiting step.

In addition to the above, the uptake of glucose, amino acids and nucleic acid precursors has been demonstrated *in vitro* for a range of adult and larval filarial worms, although the site and mechanism of transport has not been studied (Barrett, 1983).

*(c) Enzyme distribution* Correlated with the absorptive function of the cuticle, acid phosphatase and naphthylamidase activity is present in the hypodermal tissues of adult *B. pahangi*, but not in the intestine, although non-specific esterase shows the reverse distribution (Howells and Chen, 1981). High levels of acid phosphatase have also been noted in the body wall of *L. carinii*, *B. pahangi*, *D. immitis* and *Setaria* species (Yanagisawa and Koyama, 1970; Goil, Sawada and Sato, 1973; Maki and Yanagisawa, 1980a, 1980b, 1980c). In contrast, an acid protease, capable of hydrolysing haemoglobin, occurs in large amounts in the intestine of *D. immitis* (Sato, Takahashi and Sawada, 1976; Sato *et al.*, 1979; Maki, Furuhashi and Yanagisawa, 1982; Swamy and Jaffe, 1983). There is also an unconfirmed report of exogenous protease activity by the microfilariae of *O. volvulus* (Ortiz y Ortiz, Gonzalez-Barranco and Salazar Mallen, 1962).

### *3.5.5 Neurotransmitters*

Nothing is known about nerve muscle physiology in filarial worms although this is an important site of anthelmintic action in other helminths. A number of putative neurotransmitters has, however, been demonstrated. Acetylcholine has been found in the microfilariae of *Dirofilaria repens* and in adult *L. carinii* (Mellanby, 1955), whilst acetylcholine esterase has been demonstrated in *L. carinii*, *S. cervi* and *Stephanofilaria stilesi* (Bueding, 1952; Shishov, Koishibaev and Timofeeva, 1973; Rao, 1978). There is also pharmacological evidence for a cholinergic system in *B. malayi* (Hillman *et al.*, 1983).

A number of biogenic amines has been isolated from filarial worms. Histamine, noradrenalin, dopamine and serotonin have been found in adult *L. carinii*, whilst the microfilariae contained serotonin, histamine and noradrenaline (Saxena *et al.*, 1977, 1978; Rao, 1978). The adults and microfilariae of *S. cervi* contain noradrenaline, dopamine, serotonin and histamine as well as some of their metabolites, such as homovanillic acid and 5-hydroxyindole acetic acid (Rathaur and Anwar, 1979). Strangely, 4-aminobutyric acid, which appears to be the main inhibitory transmitter in nematodes, has yet to be demonstrated in filarial worms.

## 3.6 Potential targets for chemotherapy

Modern anthelmintics work on a relatively restricted number of target sites. Biochemical studies on the adult filarial worm suggest a number of new potential areas. There are, however, two problems; firstly, biochemical studies on adult female worms include a significant contribution from the microfilariae in the uterus. Second, and more important, great care has to be taken in extrapolating the results from one species of filariid to another. This is particularly so with *Litomosoides*, whose metabolism is different from *Onchocerca* species, one being an aerobe, the other a homolactic fermentor — as are most of the other filariae (see Section 3.5.1).

### *3.6.1 Nerve–muscle physiology*

Many anthelmintics effective against gut parasitic nematodes work by disrupting nerve–muscle physiology and the success of ivermectin and of the acetylcholine esterase inhibitors must make this area a prime target for chemotherapy. The site of parasitism of adult filariids within the host, however, is such that they can maintain their position without constant muscular activity. Temporary muscle paralysis is not, therefore, effective against filariids.

### *3.6.2 Metabolic inhibitors*

Studies on parasite biochemistry point to the inhibition of metabolic pathways as an obvious target for the rational development of anthelmintics. What is required is a target enzyme, the inhibition of which will be a lethal event for the parasite. Microorganisms and parasitic protozoa multiply inside their hosts by rapid asexual division. For them, inhibition of synthetic pathways is a successful control strategy because it disrupts cell multiplication. With nematodes inhibition of synthetic pathways may reduce egg or larval production, but is unlikely to be lethal to the adult, at least in the short term. Thus with filarial worms, inhibitors of synthetic pathways can lead to a suppression of microfilarial production, but would not necessarily kill the adults. If such a cessation of microfilarial production were permanent with *O. volvulus* most pathology would be avoided, but best current experimental work suggests that such effects are temporary. For this reason, inhibition of energy-producing pathways is the preferred strategy for parasitic helminths.

Parasitic helminths differ from their hosts in having an absolute requirement for carbohydrate as their sole energy source. This considerably narrows the range of possible sites for inhibition. One way of choosing target enzymes is to look for enzymes which are present in the parasite, but not in the host, so that there is a unique target. Unique enzymes can be found in the pathways of carbohydrate metabolism in helminths, but invariably they seem to occur in the terminal parts of carbohydrate breakdown, where alternative routes exist. Parasitic helminths show a great deal of flexibility in their metabolic pathways. In the trematode *Fasciola hepatica*, for example, inhibition of phosphoenolpyruvate carboxykinase by mercaptopicolinic acid merely results in a switch from propionate formation to lactate production (Lloyd and Barrett, 1983). However, a unique target enzyme may not be necessary, since there are sufficient differences in the properties of the corresponding isofunctional enzymes from nematodes and mammals to allow selective inhibition.

Of the enzymes of carbohydrate catabolism, the non-equilibrium enzymes of glycolysis (hexokinase, phosphorylase, phosphofructokinase and pyruvate kinase) must be prime target sites. Inhibition of one of the rate-limiting, non-equilibrium enzymes will result in an immediate decrease in glycolytic flux. Two factors suggest that these enzymes may be particularly suitable for selective inhibition. First, the non-equilibrium enzymes are regulatory enzymes, and as such have binding sites for a variety of effector molecules. Second, there is considerable variation in the regulatory properties of the same enzyme from different animal sources. The species-specific variation in binding sites found in regulatory enzymes should be exploitable in anthelmintic design.

In addition to modulation by effector molecules, the non-equilibrium enzymes of glycolysis may also be regulated by phosphorylation/dephosphorylation systems based on protein kinases and phosphoprotein phosphatases and by protein regulators. Interference with these systems may offer another method of attack.

### *3.6.3 Cuticle structure and function*

The easy accessibility of the cuticle to chemical agents makes it an attractive target for chemotherapy. The role of the cuticle in nutrient uptake in filariids is of special relevance in this context see Section 3.5.4(b)).

The filarial cuticle offers three possible targets: disruption of formation, alteration of surface properties and inhibition of uptake mechanisms. The final assembly of the nematode cuticle occurs extracellularly and is essentially collagenous, but unlike vertebrate collagen, the nematode cuticle is stabilized by sulphydryl links. Damage to the cuticle could render the animal more permeable, but this might upset the ionic balance of the tissues, render the parasite more vulnerable to the host's immune response and increase the penetration of anthelmintics.

## 3.7 Screening methods for detecting filaricides

The cotton rat filarial worm, *Litosomoides carinii*, was used by Hewitt *et al.* (1947) to detect the activity of DEC. It has subsequently been used to screen large numbers of compounds for filaricidal activity. However, its use is no longer supported by the WHO. One of the major disadvantages of *L. carinii* is that its microfilariae seem to be very susceptible to drugs of a wide variety of chemical groups. The literature is replete with reports of the microfilaricidal activity of compounds which turn out to have no effect on other filariae. The peculiar sensitivity of the microfilariae of *L. carinii* is probably related to the fact that it is an obligate aerobe whereas other filariae are homolactic fermentors (see Section 3.5).

If one is seeking new drugs for treating lymphatic filariasis the obvious screening systems will involve *Brugia malayi* or *B. pahangi*. Both of these parasites are relatively easily maintained in the laboratory. Of the two *B. pahangi* is rather easier to maintain but *B. malayi* has the advantage that it is a parasite of man. Both are also natural parasities of dogs and cats in which they occupy the same anatomical sites as do *B. malayi* and *W. bancrofti* in man. Clearly, dogs and cats would be far too expensive for any primary screen. Fortunately Ash and Riley (1970) showed that *Brugia* species will develop in jirds or sand rats (*Meriones unguiculatus*). McCall and coworkers (1973) showed that they will develop to full maturity in the peritoneal cavities of jirds. It is possible to screen compounds in these types of infection but an analysis of the data presented by Denham, Suswillo and Chusattayanond (1984) showed that large groups of infected jirds would be needed to ensure that a reduction in worm burden of 50 per cent could be detected in a statistically valid manner. While studying the hybridization of the *Brugia* species it was found that if adult worms produced in one jird were transplanted into the peritoneal cavities of other jirds, there was a remarkably consistent recovery of adult worms on autopsying the recipients. Suswillo and Denham (1977) developed a technique of using transplanted adult *B. pahangi* as a screen for macrofilaricidal compounds. The full details of the method have been described by Denham (1982b). Briefly, infective larvae obtained from mosquitoes are inoculated into the peritoneal cavities of jirds where they develop into adult worms. These worms are recovered and anaesthetized with methyridine to prevent them entangling themselves with the other worms and are then separated into groups of counted worms. These are implanted through a small incision into the peritoneal cavities of naive jirds. The recipient jirds are then treated with the putative macrofilaricides. In the screen operated at the London School of Hygiene and Tropical Medicine a compound is initially screened in a single jird into which has been implanted ten female and five male worms. Usually the compound is injected subcutaneously for five days at 100 mg/kg per day and autopsies are carried out 35 days later. The advantages of the transplantation system are in reduced cost, because few jirds have to be maintained for the 70 days required for full maturation of the worms and high reproducibility of the worm recoveries.

*Brugia* species also develop well in *Praomys natalensis* (the multimammate rat-mouse) when injected subcutaneously, and this system has been used by the WHO-funded team in Giessen, Federal Republic of Germany. Strangely the worms will not develop or survive in the peritoneal cavities of *Praomys*.

Secondary screening for activity against lymphatic dwelling bilariae can be carried out in cats or monkeys infected by subcutaneous inoculation of infective larvae of *B. pahangi* or *B. malayi*. The Silvered Leaf Monkey (*Presbytis cristatus*) can be infected with *W. bancrofti*, and this could represent a final screen before testing the compound in man.

Screening for antilymphatic dwelling filariae compound is, therefore, reasonably satisfactory, but unfortunately the same cannot be said for screens for anti-*Onchocerca* compounds. No *Onchocerca* species will develop to maturity in a rodent so that it is essentially impossible to screen compounds against the target organism. *Monoanema globulosa* is a filarial parasite of rodents in East Africa whose adults live in the pulmonary arteries and microfilariae live in the skin

(Muller and Nelson, 1975). However, there is no reason to think that this parasite would be a better primary screen than either *Brugia* or *Dipetalonema viteae*.

*Dipetalonema viteae* has a subcutaneously dwelling adult worm and develops in a variety of rodents. It appears to be more difficult to kill by chemotherapy than is *Brugia* and might be considered as a secondary screen for onchocecicidal drugs. Fortunately there is a tertiary screen for this purpose. Copeman (1979) has shown the utility of *Onchocerca gibsoni* infections of cattle for testing for filaricidal activity. However there are many disadvantages in using cattle in a screen—not the least of which is the amount of compound needed to treat each animal. Despite this, Copeman has screened several compounds under the sponsorship of the World Bank/UNDP/WHO Special Programme. Copeman's system involves testing the cattle and autopsying them two months later. He then collects nodules (which contain the adult worms) and performs histopathological examination of these. Skin snips are also taken from the cattle during the study period.

Assuming that the necessary toxicological tests have been made, compounds which are macrofilaricidal against *O. gibsoni* could be tested in human patients with onchocerciasis. The World Bank/UNDP/WHO Special Programme has supported two centres in West Africa where these trials can be carried out.

It would be possible to insert a final animal stage before testing compounds in man, as Duke (1962) has shown that chimpanzees (*Pan troglodites*) can be infected with *O. volvulus*. Duke (1977) has tested several potentially active compounds in chimpanzees. Most people now find it too objectionable on ethical grounds to use the chimpanzee in this way and we cannot really see any justification for inserting this extra stage into the screening process.

### 3.8 New compounds with significant antifilarial activity

Under the Special Programme scheme for supporting screening for filaricidal compounds, over 6000 compounds have been tested for activity. Screening programmes have also been run by a number of pharmaceutical companies, notably Bayer, Hoffman–LaRoche, CIBA–Geigy and Rhône Poulenc Santé. A number of compounds have shown activity against various filariae. Probably the majority of the active compounds are benzimidazole carbamates. Many of the commercially available benzimidazole carbamates are potent macrofilaricides if injected parenterally but they are not active when given *per os*. Reports of the activity of several of these compounds have been published. In terms of the dose of compounds required the two most potent benzimidazole carbamates are mebendazole (**5a**) (Denham, Suswillo and Rogers, 1978) and flubendazole (**5b**) (Denham *et al.*, 1979), the latter being marginally better. Mebendazole has a CD95 (i.e. the dose needed to kill 95 per cent of the worms) of 5 × 6.3 mg/kg in the jird *Brugia* transplant test and flubendazole a CD95 of 5 × 3.2 mg/kg. Flubendazole is also a potent chemoprophylatic agent against *B. pahangi* (Chusattayanond and Denham, 1984) when injected subcutaneously, and because of its depot effect it can protect jirds for over three months.

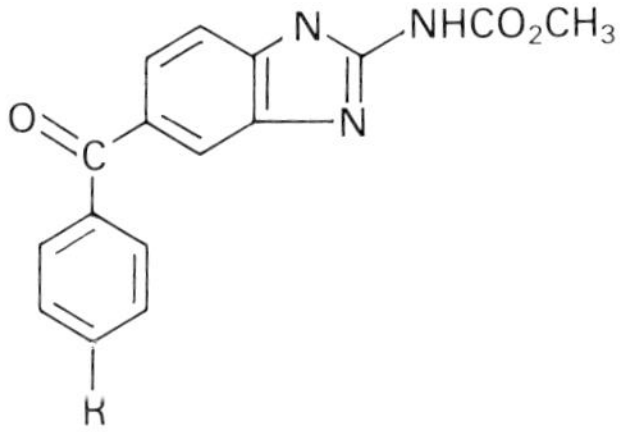

(**5a**) Mebendazole, R = H
(**5b**) Flubendazole, R = F

Unfortunately these benzimidazole carbamates were not macrofilaricidal in Copeman's *O. gibsoni* test. They were embryotoxic to the worms but the duration of this embryostatic effect has not been determined. This effect has also been demonstrated in human onchocerciasis (Rivas-Alcala *et al.*, 1984.)

Mebendazole is active against *W. bancrofti* in humans when given *per os* for 14 days (Narasimham *et al.*, 1978). However, one only expects dramatic effects with this class of compound if they are given by injection, and a trial with one form of injectable flubendazole showed activity against *O. volvulus* but painful sterile abscesses were produced at the injection site.

## 3.9 Possible mode(s) of action of filaricides

It is difficult to pinpoint the site of action of anthelmintics known to be active against filarial worms. Most of them probably act on several targets and a report claiming that an anthelmintic inhibits a particular enzyme *in vitro* is no surety that this is the critical target *in vivo*.

The mode of action of the two major antifilarial drugs, suramin and DEC is unknown. Suramin (**2**) has been reported to inhibit a wide range of enzymes including lactate dehydrogenase, malate dehydrogenase, malic enzyme, dihydrofolate reductase, 10-formyltetratydrofolate dehydrogenase, succinic dehydrogenase, glyceraldehyde-3-phosphate dehydrogenase and protein kinase (Jaffe, 1972, 1980a; Walter and Schulz-Key, 1980a, 1980b, 1980c; Walter and Vanden Bossche, 1980; Walter and Albiez, 1981). However, with whole worms suramin has no effect on glycolysis either *in vivo* or *in vitro*. The evidence suggests that *in vivo* suramin damages the intestinal mucosa of the worm (Howells, Mendis and Bray, 1983).

A number of mechanisms may combine to produce the observed effects of DEC (**1**). The piperazine ring of DEC probably acts on the neuromuscular system to produce a reversible paralysis. However, unlike piperazine this is often preceded by increased activity (Sanderson, 1970). The other main action of DEC on microfilariae may be to unmask, by an unknown mechanism, previously hidden antigenic determinants on the cuticle so that the parasites are destroyed by the host's immune response (Piessens and Beldekas, 1979). A number of enzymes have also been reported to be inhibited by DEC including NADP-independent

10-formyltetrahydrofolate dehydrogenase, pyrophosphatase and acetylcholine esterase (Sanderson, 1970; Jaffe, 1980a; Walter and Schulz-Key, 1981).

Organophosphorus anthelmintics such as metrifonate (**3**) inhibit acetylcholine esterase. The acetylcholine esterases of nematodes (and insects) are much more sensitive to inhibition by organophosphorus compounds than is the corresponding mammalian enzyme. The nematode cholinesterase–organophosphorus complex is more stable than that of mammals, correlating with the known relative toxicity of these compounds (Sanderson, 1970).

Amoscanate, *N*-(4-isothiocyanatophenyl)-4-nitroaniline (**6**), depresses glucose uptake and transport in *L. carinii* and *B. pahangi* (Nelson and Saz, 1984), but the mechanism is unknown.

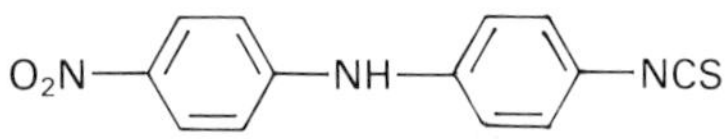

$O_2N$—⟨benzene⟩—NH—⟨benzene⟩—NCS

(**6**) Amoscanate

Arsenical compounds react with thiol groups and are potent inhibitors of a wide range of enzymes including cholinesterases, 2-oxoacid oxidases, ATPases and kinases. The trivalent organic arsenical Mel-W, therefore, has many possible sites of action. The phosphofructokinases of filarial worms are considerably more sensitive to inhibition by trivalent organic antimonials than is the mammalian enzyme (Section 3.5); this could also be a site of action for organic arsenicals.

The benzimidazole carbamates (mebendazole, cambendazole, flubendazole, albendazole, oxibendazole, fenbendazole, oxfendazole) probably all work in the same basic way, but variations in their pharmokinetic properties may account for the different effects observed.

A variety of effects of benzimidazoles on bacteria and fungi have been suggested, including incorporation into nucleic acids, disruption of electron transport and interference with vitamin $B_{12}$ synthesis or usage (Allen and Gottlieb, 1970; Stutzenberger and Parle, 1973). There are reports of benzimidazoles inhibiting electron transport and fumarate reductase activity in nematodes (Prichard, 1973; Köhler and Bachmann, 1978) and causing increases in acetylcholine esterase levels in trichostronglyes (Rapson, Lee and Watts, 1981). However, the primary action of benzimidazoles is to bind to unpolymerized tubulin, preventing microtubule formation (Friedman and Platzer, 1978). The resulting cellular disruption leads to inhibition of transport mechanisms, impairment of synthesis and finally cytoplasmic autolysis and cellular necrosis. Disruption of microtubule formation also accounts for the ovicidal effects of benzimidazoles (Coles, 1977). There is also evidence that benzimidazoles may be actively taken up by nematodes (Coles, 1977). Avermectins are semi synthetic derivatives of macrocylic lactones produced by the mycelia of *Streptomyces avermilitis*. Ivermectin (**4**), the commercially available avermectin, has a broad spectrum of activity against nematodes and parasitic arthropods and is effective at extremely low therapeutic doses, 10 or 100 μg/kg. Ivermectin disrupts transmission at γ-aminobutyric acid (GABA)-mediated synapses. In mammals, GABA-mediated synapses occur principally in the central nervous system, whilst in invertebrates they are important transmitters in the peripheral nervous system. The mammalian GABA synapse is also disrupted by ivermectin, but at the therapeutic doses used ivermectin does not cross the blood brain barrier.

In the nematode *Ascaris lumbricoides* ivermectin blocks the transmission of nervous impulses from the ventral interneurones to the motor neurones (Kass *et al.*, 1982), probably via increased GABA action (GABA being an inhibitory neurotransmitter in nematodes). Ivermectin enhances the GABA effect through two closely related mechanisms: first, it potentiates the release of GABA from presynaptic vesicles and, second, increases the binding of GABA to receptor sites (Kass *et al.*, 1980; Pong and Wang, 1980; Pong, Wang and Fritz, 1980). There have also been suggestions that, at least in mammalian systems, ivermectin may have both GABA agonist and antagonist properties (Paul, Skolnick and Zatz, 1980).

There have, so far, been no claims for ivermectins acting as enzyme inhibitors, although such reports will probably eventually appear. The basis of the microfilaricidal effect observed with ivermectins is uncertain. It may be the result of paralysis either of the microfilariae or of the uterine muscles, preventing the escape of the microfilariae from the uterus, or the ivermectins may have an, as yet, undiscovered metabolic effect.

Nothing is known about the mode of action of some of the more recent compounds which have shown promise as filaricides, drugs such as desmethylmisonidazole (**7**), furapyrimidone (**8**) and the CIBA–Geigy compounds CGP 6140 (**9**), CGP 20376 (**10a**) and CGP 24914 (**10b**).

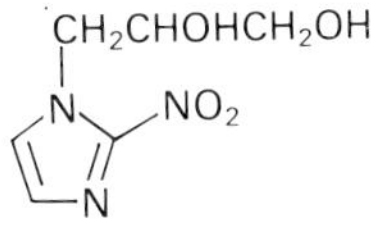

$CH_2CHOHCH_2OH$, N, $NO_2$, N

(**7**) Desmethyl misnidazole

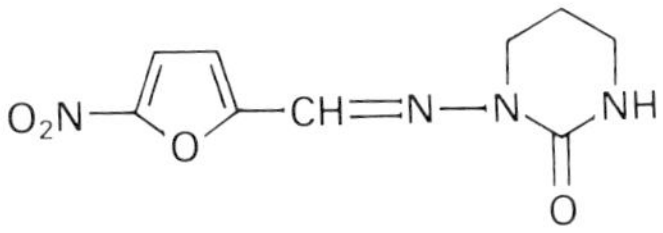

$O_2N$, O, CH=N—N, NH, O

(**8**) Furapyrimidone

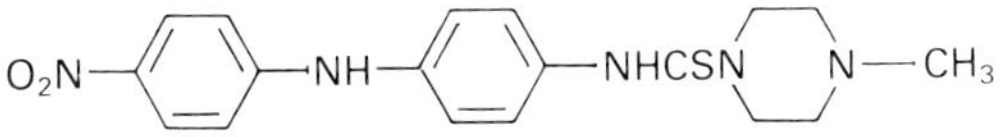

$O_2N$—, NH—, —NHCSN, N—$CH_3$

(**9**) CGP 6140

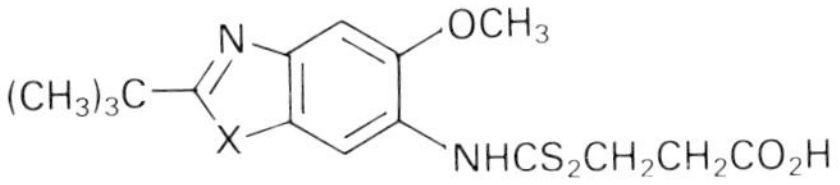

N, $OCH_3$, $(CH_3)_3C$—, X, $NHCS_2CH_2CH_2CO_2H$

(**10a**) CGP 20376, X = S
(**10b**) CGP 24914, X = O

It is difficult to know what the actual tissue levels of anthelmintics in the region of the parasite might be *in vivo*, paricularly for filarial worms. This makes correlations between *in vitro* and *in vivo* studies difficult. If the anthelmintics were evenly distributed throughout the host, under normal does regimens the concentration would be about $10^{-4}M$ and $10^{-6}M$ for ivermectins. Some anthelmintics, such as the benzimidazoles, may be actively accumulated by nematodes, resulting in higher tissue levels in the parasite than the host.

The biochemical mode of action of anthelmintics is a rapidly expanding field. However, until there is a better understanding of the pharmokinetics of anthelmintics within the host–parasite complex a complete explanation of how and why anthelmintics work will not be possible.

### 3.10 References


1. A.I. Akinwande, and E.O. Akinrimisi, (1980). *IRCS Med. Sci.*, **8**, 565–566.
2. P.M. Allen, and D. Gottlieb, (1970). *Appl. Microbiol.*, **20**, 919–926.
3. N. Anwar, A.A. Ansari, and S. Ghatak, (1975). *Proc. Ind. Nat. Sci. Acad.*, **41B**, 550–558.
4. N. Anwar, A.A. Ansari, S. Ghatak, and C.R. Krishna Murti, (1977a). *Z. fur Parasitkde*, **51**, 275–283.
5. N. Anwar, R.K. Chatterjee, A.B. Sen, and S.N. Ghatak, (1977b). *Z. fur Parasitkde*, **54**, 79–82.
6. N. Anwar, A.K. Srivastava, and S. Ghatak, (1978). *Ind. J. Parasitol.*, **2**, 101–105.
7. L.R. Ash, and J.H. Riley, (1970). *J. Parasitol.*, **56**, 962–968.
8. A. Awadzi, K.Y. Dadzie, H. Schulz-Key, D.R.W. Haddock, H.M. Gilles, and M.A. Aziz, (1984). *Lancet*, **ii**, 921.
9. K. Awadzi, and H.M. Gilles, (1980). *Ann. Trop. Med. Parasitol.*, **74**, 199–210.
10. M.A. Aziz, S. Diallo, I.M. Diop, M. Larivere, and M. Porta, (1982). *Lancet*, **ii**, 171–173.
11. J. Barrett, (1981). *Biochemistry of Parasitic Helminths*, Macmillan, London.
12. J. Barrett, (1983). *Helminth. Abstr.*, **A52**, 1–18.
13. J. Barrett, (1984). *Parasitology*, **88**, 179–198.
14. J. Barrett, and P.E. Butterworth, (1984). *Biochem. J.*, **221**, 535–540.
15. S. Berls, and E. Bueding, (1951). *J. Biol. Chem.*, **191**, 401–418.
16. J.B. Brazier, and J.J. Jaffe, (1973). *Comp. Biochem. Physiol.*, **44B**, 145–155.
17. E. Bueding, (1949a). *J. Expl Med.*, **89**, 107–130.
18. E. Bueding, (1949b). *Physiol. Rev.*, **29**, 195–218.
19. E. Bueding, (1949c). *Fed. Proc.*, **8**, 188–189.
20. E. Bueding, (1952). *Br. J. Pharmac. Chemother.*, **7**, 563–566.
21. E. Bueding, and B. Charms, (1951). *Nature (Lond.)*, **167**, 149.
22. E. Bueding, and B. Charms, (1952). *J. Biol. Chem.*, **196**, 615–627.
23. S.N. Chen, and R.E. Howells, (1979a). *Parasitology*, **78**, 343–354.
24. S.N. Chen, and R.E. Howells, (1979b). *Ann. Trop. Med. Parasitol.*, **73**, 473–486.
25. S.N. Chen, and R.E. Howells, (1979c). *Expl Parasitol.*, **47**, 209–221.
26. S.N. Chen, and R.E. Howells, (1981a). *Expl Parasitol.*, **51**, 296–306.
27. S.N. Chen, and R.E. Howells, (1981b). *Ann. Trop. Med. Parasitol.*, **75**, 329–334.
28. P.V. Cherian, B.E. Stromberg, D.J. Weiner, and E.J.L. Soulsby, (1980). *Int. J. Parasitol.*, **10**, 227–233.
29. W. Chusattayanond, and D.A. Denham, (1984). *J. Parasitol.*, **70**, 191–192.
30. G.C. Coles, (1977). *Pesticide Sci.*, **8**, 536–543.
31. J.C.W. Comley, and J.J. Jaffe, (1981). *J. Parasitol.*, **67**, 609–616.
32. J.C.W. Comley, and J.J. Jaffe, (1983). *Biochem. J.*, **214**, 367–376.
33. J.C.W. Comley, J.J. Jaffe, and L.R. Chrin, (1982). *Mol. Biochem. Parasitol.*, **5**, 19–31.
34. J.C.W. Comley, J.J. Jaffe, L.R. Chrin, and R.B. Smith, (1981). *Mol. Biochem. Parasitol.*, **2**, 271–283.
35. D.B. Copeman, (1979). *Tropenmed. Parasitol.*, **30**, 469–474.
36. J.E. Deas, F.J. Aguilar, and J.H. Miller, (1974). *J. Parasitol.*, **60**, 1006–1012.

37. D.A. Denham, and P. Mellor, (1976). *J. Helminth.*, **50**, 49–52.
38. D.A. Denham, R. Samad, S-Y. Cho, R.R. Suswillo, and S.C. Skippins, (1979). *Trans. R. Soc. Trop. Med. Hyg.*, **73**, 673–676.
39. D.A. Denham, R.R. Suswillo, and W. Chusattayanond, (1984). *Parasitology*, **88**, 295–301.
40. D.A. Denham, R.R. Suswillo, and R. Rogers, (1978). *Trans. R. Soc. Trop. Med. Hyg.*, **72**, 546–647.
41. B.O.L. Duke, (1962). *Trans. R. Soc. Trop. Med. Hyg.*, **56**, 271.
42. B.O.L. Duke, (1968). *Bull. WHO*, **39**, 157–167.
43. B.O.L. Duke, (1970). *Bull. WHO*, **42**, 115–117.
44. B.O.L. Duke, (1977). *Troponmed. Parasitol.*, **28**, 447–455.
45. P.R. Earle, (1959). *Ann. N.Y. Acad. Sci.*, **77**, 163–175.
46. H.A. Flockhart, and D.A. Denham, (1984). *J. Parasitol.*, **70**, 378–384.
47. P.A. Friedman, and E.G. Platzer, (1978). *Biochem. Biophys. Acta*, **544**, 605–614.
48. A. Furman, and L.R. Ash, (1983). *J. Parasitol.*, **69**, 1043–1047.
49. M.M. Goil, T. Sawada, and K. Sato, (1973). *Jap. J. Expl Med.*, **43**, 215–218.
50. S.L. Govindwar, T.S. Gawande, and B.C. Harinath, (1974). *Ind. J. Biochem. Biophys.*, **11**, 338–339.
51. F. Hawking, and D.A. Denham, (1976). *Trop. Dis. Bull.*, **73**, 348–373.
52. H. Hayashi, and H. Oya, (1978). Fourth International Congress on Parasitology, Warsaw, Abstract F, pp. 73–74.
53. R.I. Hewitt, S. Kushner, H.W. Stewart, E. White, W.S. Wallace, and Y. Subbarow, (1947). *J. Lab. Clin. Med.*, **32**, 1314–1329.
54. G.R. Hillman, A. Ewert, L. Westerfield, and S.O. Grim, (1983). *Comp. Biochem. Physiol.*, **74C**, 299–301.
55. R.E. Howells, (1980). In *The Host–Invader Interplay* (Ed. H. Vanden Bossche), pp. 69–84, Elsevier, Amsterdam.
56. R.E. Howells, and S.N. Chen, (1981). *Expl Parasitol.*, **51**, 42–58.
57. R.E. Howells, A.M. Mendis, and P.G. Bray, (1983). *Parasitology*, **87**, 29–48.
58. W.F. Hutchison, and K.M. McNeill, (1970). *Comp. Biochem. Physiol.*, **35**, 721–727.
59. W.F. Hutchison, F.J. Oelshlegel, D. Sullivan, and A.C. Turner, (1978). *J. Mississippi Acad. Sci.*, **23** (Suppl.), 50.
60. W.F. Hutchison, and A.C. Turner, (1979a). *Comp. Biochem. Physiol.*, **64B**, 399–401.
61. W.F. Hutchison, and A.C. Turner, (1979b). *Comp. Biochem. Physiol.*, **62B**, 71–73.
62. W.F. Hutchison, A.C. Turner, and F.J. Oelshlegel, (1977). *Comp. Biochem. Physiol.*, **58B**, 131–134.
63. J.J. Jaffe, (1971). *Ann. N.Y. Acad. Sci.*, **186**, 113–114.
64. J.J. Jaffe, (1972). In *Comparative Biochemistry of Parasites* (Ed. H. Vanden Bossche), pp. 219–233, Academic Press, New York.
65. J.J. Jaffe, (1980a). In *The Host–Invader Interplay* (Ed. H. Vanden Bossche), pp. 605–614, Elsevier, Amsterdam.
66. J.J. Jaffe, (1980b). In *Report of the Fifth Meeting of the Scientific Working Group on Filariasis*, pp. 19–20, WHO, Geneva.
67. J.J. Jaffe, and L.R. Chrin, (1978a). *J. Parasitol.*, **64**, 661–668.
68. J.J. Jaffe, and L.R. Chrin, (1978b). *J. Parasitol.*, **64**, 769–774.
69. J.J. Jaffe, and L.R. Chrin, (1979a). *J. Parasitol.*, **65**, 550–554.
70. J.J. Jaffe, and L.R. Chrin, (1979b). *J. Parasitol.*, **65**, 226–232.
71. J.J. Jaffe, and L.R. Chrin, (1979c). *Biochem. Pharmac.*, **28**, 1831–1835.
72. J.J. Jaffe, and L.R. Chrin, (1980). *J. Parasitol.*, **66**, 53–58.
73. J.J. Jaffe, and L.R. Chrin, (1981). *Mol. Biochem. Parasitol.*, **2**, 259–270.
74. J.J. Jaffe, and H.M. Doremus, (1970). *J. Parasitol.*, **56**, 254–260.
75. J.J. Jaffe, J.J. McCormack, and E. Meymarian, E. (1972). *Biochem. Pharmac.*, **21**, 719–731.
76. J.J. Jaffe, L.R. Chrin, and R.B. Smith, (1980). *J. Parasitol.*, **66**, 428–433.
77. J.J. Jaffe, J.C.W Comley, and L.R. Chrin, (1982). *Mol. Biochem. Parasitol.*, **5**, 361–370.
78. J.J. Jaffe, J.J. McCormack, E. Meymarian, and H.M. Doremus, (1977). *J. Parasitol.*, **63**, 547–553.
79. K.H. Johnson, and W.J. Bemrick, (1969). *Am. J. Vet. Res.*, **30**, 1443–1450.
80. I.S. Kass, D.A. Larsen, C.C. Wang, and A.O.W. Stretton, (1982). *Expl Parasitol.*, **54**, 166–174.
81. I.S. Kass, C.C. Wang, J.P. Walrond, and A.O.W. Stretton, (1980). *Proc. Natn. Acad. Sci. USA*, **77**, 6211–6215.

82. H. Khatoon, Wajihullah, and A. Ansari, (1983). *Helminthologia*, **20**, 215–220.
83. P. Köhler, and R. Bachmann, (1978). *Mol. Pharmacol.*, **14**, 155–163.
84. P.R. Komuniecki, and H.J. Saz, (1982). *J. Parasitol.*, **68**, 221–227.
85. B.W. Langer, and D. Jiampermpoon, (1970). *J. Parasitol.*, **56**, 144–145.
86. C-C. Lee, and J.H. Miller, (1967). *Exp. Parasitol.*, **20**, 334–344.
87. C.-C. Lee, and J.H. Miller, (1969). *J. Parasitol.*, **55**, 1035–1045.
88. G.M. Lloyd, and J. Barrett, (1983). *Expl Parasitol.*, **56**, 259–265.
89. J.W. McCall, J.B. Malone, H.S. Ah, and P.E. Thompson, (1973). *J. Parasitol.*, **59**, 436.
90. K.M. McNeill, and W.F. Hutchison, (1971). *Comp. Biochem. Physiol.*, **38B**, 493–500.
91. K.M. McNeill, and W.F. Hutchison, (1972). In *Canine Heartworm Disease: The Curren Knowledge* (Ed. R.E. Bradley), pp. 51–55, University of Florida, Gainesville.
92. J.W. Mak, and V. Zaman, (1980). *Trans. R. Soc. Med. Hyg.*, **74**, 285–291.
93. J. Maki, A. Furuhashi, and T. Yanagisawa, (1982). *Parasitology*, **84**, 137–147.
94. J. Maki, and T. Yanagisawa, (1980a). *J. Helminth.*, **54**, 39–41.
95. J. Maki, and T. Yanagisawa, (1980b). *Parasitology*, **80**, 23–28.
96. J. Maki, and T. Yanagisawa, (1980c). *Parasitology*, **81**, 603–608.
97. L. Mazzoti, (1948). *Rev. Inst. Salubridad Enfermed. Trop.*, **9**, 235–237.
98. H. Mellanby, (1955). *Parasitology*, **45**, 287–294.
99. K.R. Middleton, and H.J. Saz, (1979). *J. Parasitol.*, **65**, 1–7.
100. R.L. Muller, and G.S. Nelson, (1975). *J. Parasitol.*, **61**, 606–609.
101. M.V.V.L. Narasimham, S.P. Roychowdhury, M. Das, and C.K. Rao, (1978). *S.E. Asian J. Trop. Med. Pub. Hlth*, **9**, 571–575.
102. H. Nduka, and R.E. Howells, (1980). Proceedings of Third European Multicolloquium o Parasitology, Cambridge, p. 10.
103. N.F. Nelson, and H.J. Saz, (1982). *J. Parasitol.*, **68**, 1162–1163.
104. N.F. Nelson, and H.J. Saz, (1984). *J. Parasitol.*, **70**, 194–195.
105. P. Oothuman, D.M. Moss, and S.E. Maddison, (1984). *J. Parasitol.*, **69**, 994–996.
106. L. Ortiz y Ortiz, D. Gonzalez-Barranco, and M. Salazar Mallen, (1962). *Salud Publica de Mexico*, **4**, 1075–1077.
107. E.A. Ottesen, (1984). WHO/FIL/84.174 pp. 1–24 (WHO document described as 'not a forma publication' but can be obtained from Geneva on request).
108. G.T. Pandya, (1961). *Z. fur Parasitkde*, **20**, 466–469.
109. F. Partono, Purnomo, and A. Soewarta, (1979). *Trans. R. Soc. Trop. Med. Hyg.*, **73**, 536–542.
110. S.M. Paul, P. Skolnick, and M. Zatz, (1980). *Biochem. Biophys. Res. Comm.*, **96**, 632–638.
111. L. Peters, E. Bueding, A. Valk, A. Higashi, and A.D. Welch, (1949). *J. Pharmac. Expl Ther.*, **95**, 212–239.
112. M. Philipp, and F.D. Rumjaneck, (1984). *Mol. Biochem. Parasitol.*, **10**, 245–268.
113. W.F. Piessens, and M. Beldekas, (1979). *Nature (Lond.)*, **282**, 845–847.
114. S.-S. Pong, and C.C. Wang, (1980). *Abstr. Soc. Neurosci.*, **6**, 184.22.
115. S.-S. Pong, C.C. Wang, and L.C. Fritz, (1980). *J. Neurochem.*, **34**, 351–358.
116. R.K. Prichard, (1973). *Int. J. Parasitol.*, **3**, 409–417.
117. T. Ramp, and P. Köhler, (1982). Abstract of the Fifth International Congress on Parasitology, Toronto, Canada, *Mol. Biochem. Parasitol.* (Suppl.), p. 141.
118. K.N. Rao, (1978). *Comp. Physiol. Ecol.*, **3**, 256–257.
119. E.B. Rapson, D.L. Lee, and S.D.M. Watts, (1981). *Mol. Biochem. Parasitol.*, **4**, 9–15.
120. S. Rathaur, and N. Anwar, (1979). *Ind. J. Biochem. Biophys.*, **16** (Suppl.), 14.
121. S. Rathaur, N. Anwar, R.K. Chatterjee, A.B. Sen, and S. Ghatak, (1980). *Ind. J. Parasitol.*, **4**, 67–69.
122. S. Rathaur, N. Anwar, J.K. Saxena, and S. Ghatak, (1982). *Z. fur Parasitdke*, **68**, 331–338.
123. R.S. Rew, and H.J. Saz, (1977). *J. Parasitol.*, **63**, 123–129.
124. R. Rivas-Alcala, C.D. Mackenzie, E. Gomez-Rojo, B.M. Greene, and H.R. Taylor, (1984). *Tropenmed. Parasitol.*, **35**, 71–77.
125. M. Salazar-Mallen, D. Gonzalez-Barranco, and H. Del Carmen Moutes, (1971). *Rev. Inst. Med. Trop. Sao Paulo*, **13**, 363–368.
126. B.E. Sanderson, (1970). *Comp. Gen. Pharmac.*, **1**, 135–151.
127. D. Santiago-Stevenson, J. Oliver-Gonzalez, and R.I. Hewitt, (1947). *J. Am. Med. Ass.*, **135**, 708–712.
128. K. Sato, M. Suzuki, S. Ohshiro, and Y. Nakagawa, (1979). *Jap. J. Parasitol.*, **28**, 49.
129. K. Sato, J. Takahashi, and T. Sawada, (1976). *Jap. J. Parasitol.*, **25** (Suppl.), 8–9.

130. J.K. Saxena, R. Sen, R.K. Chatterjee, and S. Ghatak, (1978). Abstract of the Asian Congress on Parasitology, Bombay, 1978, p. 113.
131. B.A. Shishov, G. Koishibaev, and T.N. Timofeeva, (1973). *Trudy Gel'mintologicheskoi Laboratorii (Ekologiya i taksonomiya gel'mintov)*, **23**, 203–211.
132. G. Singh, N.A. Pampori, and V.M.L. Srivastava, (1984). *Ind. J. Expl Biol.*, **22**, 50–53.
133. V.M.L. Srivastava, R.K. Chatterjee, A.B. Sen, S. Ghatak, and C.R. Krishna Murti, (1970). *Expl Parasitol.*, **28**, 176–185.
134. V.M.L. Srivastava, and S. Ghatak, (1971). *Ind. J. Biochem. Biophys.*, **8**, 108–111.
135. V.M.L. Srivastava, and S. Ghatak, (1974). *Ind. J. Expl Biol.*, **12**, 472–473.
136. V.M.L. Srivastava, S. Ghatak, and C.R. Krishna Murti, (1968). *Expl Parasitol.*, **23**, 339–346.
137. V.M.L. Srivastava, S. Ghatak, and C.R. Krishna Murti, (1970). Second International Convention of Biochemistry, Baroda (India), Abstract A 79, p. 28.
138. L. Stryer, (1981). *Biochemistry*, 2nd ed., Freeman, San Francisco.
139. D. Sturchler, F. Wyss, and A. Hank, (1981). *Trans. R. Soc. Trop. Med. Hyg.*, **75**, 617.
140. F.J. Stutzenberger, and N.J. Parle, (1973). *J. Gen. Microbiol.*, **76**, 197–209.
141. D. Subrahmanyam, (1967). *Can. J. Biochem.*, **45**, 1195–1197.
142. R.R. Suswillo, and D.A. Denham, (1977). *J. Parasitol.*, **63**, 591–592.
143. K.H.S. Swamy, and J.J. Jaffe, (1983). *Mol. Biochem. Parasit.*, **9**, 1–14.
144. G.C. Thooris, (1956). *Bull. Soc. Path. Exot.*, **49**, 1138–1157.
145. A.C. Turner, and W.F. Hutchison, (1979). *Comp. Biochem. Physiol.*, **64B**, 403–405.
146. A.C. Turner, and W.F. Hutchison, (1983). *Comp. Biochem. Physiol.*, **73B**, 331–334.
147. L. van Hoof, C. Henrard, E. Peel, and M. Wanson, (1947). *Ann. Soc. Belge Med. Trop.*, **27**, 172–177.
148. T. von Brand, (1960). In *Nematology* (Eds. J.N. Sasser and W.R. Jenkins), pp. 233–241, University of North Carolina Press, Chapel Hill.
149. T. von Brand, I.B.R. Bowman, P.P. Weinstein, and T.K. Sawyer, (1963). *Expl Parasitol.*, **13**, 128–133.
150. R.D. Walter, and E.J. Albiez, (1981). *Mol. Biochem. Parasitol.*, **4**, 53–60.
151. R.D. Walter, and H. Schulz-Key, (1980a). In *The Host–Invader Interplay* (Ed. H. Vanden Bossche), pp. 709–712, Elsevier, Amsterdam.
152. R.D. Walter, and H. Schulz-Key, (1980b). *Zbl. Bakt. Hyg. I Abt. Ref. BD*, **267**, 314.
153. R.D. Walter, and H. Schulz-Key, (1980c). *Tropenmed. Parasitol.*, **31**, 55–58.
154. R.D. Walter, and H. Schulz-Key, (1981). In *The Biochemistry of Parasites* (Ed. G.M. Slutzky), p. 223, Pergamon, Oxford.
155. R.D. Walter, and H. Vanden Bossche, (1980). Report of the Fifth Meeting of the Scientific Working Group on Filariasis, p. 21, WHO, Geneva.
156. E.J. Wang, and H.J. Saz, (1974). *J. Parasitol.*, **60**, 316–321.
157. A.D. Welch, L. Peters, E. Bueding, A. Valk, and A. Higashi, (1947). *Science*, **105**, 486–488.
158. T. Wilson, (1950). *Trans. R. Soc. Trop. Med. Hyg.*, **44**, 49–66.
159. T. Yanagisawa, and T. Koyama, (1970). The Joint Conference of Parasitic Diseases. The United States–Japan Cooperative Medical Science Program, pp. 22–23.
160. T. Yonezawa, (1952). *Igaku to Seibutsugaku (Med. Biol.)*, **25**, 149–152.
161. T. Yonezawa, (1953). *Med. Bull. Kagoshimi Univ.*, December **1953**, 4–6.

# 4 Trypanosomiasis and leishmaniasis

J.R. Brown
Tropical Diseases Chemotherapy Research Unit and Department of Pharmaceutical Chemistry, Sunderland Polytechnic, Sunderland SR2 7EE, UK



## 4.1 Introduction

Those who live in the Western World do not easily comprehend the scale of the problems posed, in tropical countries, by malnutrition and disease. The former has been highlighted by the appalling famine in Ethiopia, yet even without disasters of this magnitude the problem of malnutrition is immense. For example, the estimated worldwide child death toll from this cause in 1981 was 17 million — equivalent to about one-third of the population of the United Kingdom. Diseases caused by parasitic protozoa are estimated to affect about a quarter of the world population at any one time. These diseases include malaria, leishmaniasis, trypanosomiasis, babesiasis, giardiasis and trichomoniasis; the first three are included within the six diseases identified by the UNDP/World Bank/WHO Special Programme for Research and Training in Tropical Diseases as major world health problems. Malaria is the most widespread, with over 600 million people at risk. Whilst antimalarial drugs are available, drug resistance is becoming a major problem: chloroquine-resistant *Plasmodium falciparum* is widespread in South East Asia and Central America, and has now gained a toehold in East Africa. The situation with respect to the various forms of trypanosomiasis and leishmaniasis is worse, since drugs are either not available or are inadequate. This review will consider trypanosomiasis and leishmaniasis, the limitations of the existing drugs for treatment of these diseases, the sites of action of these drugs and possible targets and strategies for new drug design. We need to answer the question: 'Is drug treatment for trypanosomiasis and leishmaniasis ineffective due to an inherent difficulty in selective killing of the organisms, or to insufficient investment in this area of research, or to a misdirection of emphasis in research on these diseases?' To make a start towards answering these questions, we need to consider first the causative organisms and the nature of these diseases.

Both trypanosomiasis and leishmaniasis are caused by protozoa of the order Kinetoplastida, so called because in addition to nuclear DNA there is a 'kinetoplast' of catenated maxi- and minicircles of DNA, located in the single large mitochondrion of these organisms. All the parasitic protozoa considered in this review have an insect vector, i.e. part of the life cycle of the organism occurs in man and the remainder in an insect. The organism adopts different morphological forms in man and in the insect vector; indeed it may adopt different morphological forms in different sites in man. The diseases to be considered here are all zoonoses, i.e. in addition to man, the organisms each parasitize specific animals, which thus act as a reservoir of infection.

## 4.2 Trypanosomiasis

Trypanosomiases are diseases caused by protozoa of the genus *Trypanosoma* and there are two major forms occurring in man. First, there is African trypanosomiasis which has a salivarian mode of transmission, i.e. the organism is deposited in the saliva of the insect vector (tsetse fly) when it takes a blood meal. Second, there is South American trypanosomiasis (Chagas' disease) which has a stercorarian mode of transmission, i.e. the organism is deposited in the faeces of the insect vector (reduviid bugs). The insect defaecates after it has taken a blood meal and the parasites then enter the body by being inadvertently rubbed into abrasions, e.g. the puncture caused by the bug itself. African trypanosomiasis occurs in the tsetse belt of equatorial Africa: 50 million people are at risk but due to careful monitoring and control, the incidence is remarkably low, with 10 000 to 20 000 cases per annum, but any breakdown of control leads to a flare-up of the disease. The disease is termed sleeping sickness from the symptoms manifested in the late stages when organisms have invaded the central nervous system. When this stage is reached the disease is second only to rabies in lethality. There is virtually 100 per cent fatality if no treatment is given. There are also animal trypanosomiases related to the human disease, the most significant being ngana in cattle: this disease renders 7 million $km^2$ (an area greater than that of the United States) of sub-Saharan Africa of little use for beef production. Ormerod (1979) has argued that elimination of animal trypanosomiasis from sub-Saharan Africa will lead to impoverishment of the land, though it is generally felt that meat production in this area could be at least doubled. Several drugs are currently used to obtain some control of ngana but the situation is deteriorating as resistance is developing (Leach and Roberts, 1981).

Human sleeping sickness and ngana in cattle are caused by three species of *Trypanosoma brucei*. *T. brucei brucei* is one causative organism of ngana in cattle. *Trypanosoma brucei gambiense* is responsible for sleeping sickness in West Africa (gambian sleeping sickness) and *T. brucei rhodesiense* causes rhodesian sleeping sickness in East Africa. This is more acute than the gambian form, death occurring within a few months. The human diseases and their chemotherapy have been reviewed recently by de Raadt (1976), Apted (1980), Evans (1981), Foulkes (1981), Brown (1983), Molyneux and Ashford (1983) and Meshnick (1984a). On taking a blood meal the infected tsetse deposits 'metacyclic' forms of the organism which transform to the long slender trypomastigote morphological form which is adapted for life in the bloodstream of the host (man). The trypomastigotes multiply but the immune system soon eliminates all the organisms except those that have switched their surface antigen (see Section 4.6). Hence there are waves of parasite numbers in the blood. As the disease progresses, the trypomastigotes also invade the CNS, being found in the cerebrospinal fluid; possibly the organisms enter and divide in the ependymal cells of the choroid plexus (these constituting the blood–brain barrier at this point) (Abolarin *et al.*, 1983). Some of the long slender bloodstream trypomastigote forms transform to a 'short stumpy' trypomastigote form which is infective for the tsetse fly; a tsetse taking a blood meal will thus become infected, and so the life cycle continues. Whilst there is effective drug treatment for sleeping sickness the drugs are far from ideal in that they must be used in the hospital setting and also have unacceptable toxicity. The drug of choice for the early stage of the disease is suramin (**1**)—a drug introduced in the 1920s following a line of development directly stemming from Ehrlich's work. It gives virtually 100 per cent cures in the early stage of the disease, but is ineffective against trypanosomes in the CNS. However, it has to be given intravenously (i.v.) and commonly causes vomiting, pruritis and (more importantly) nephrotoxicity. Moreover, 1 in 3000 persons are hypersensitive to the drug. Pentamidine (**2**) is also used in early stages of the disease, but only as a second-line drug (since it is not effective against the rhodesian form) and as a prophylactic (in West Africa). It was introduced in the 1940s. It must be given i.v. and often causes vomiting, headaches, dizziness and hypotension due to histamine release. The drug of choice in late-stage sleeping sickness, when the CNS is invaded, is melarsoprol (**3**). It was introduced in the 1950s and is given i.v. This drug causes an encephalopathy which is fatal in up to 5 per cent of the cases treated. Nifurtimox (**4**) is used as a last resort if melarsoprol is ineffective. Clearly new drugs are urgently required.

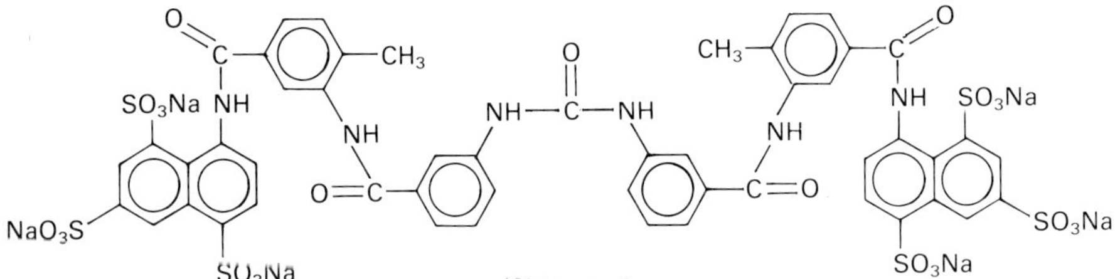

(**1**) Suramin

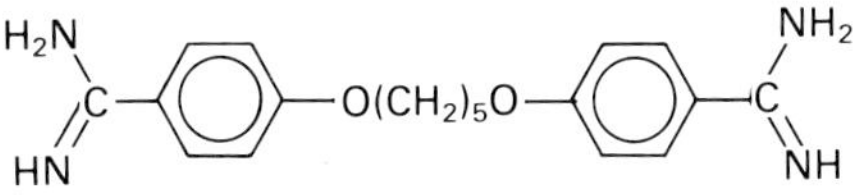

(**2**) Pentamidine

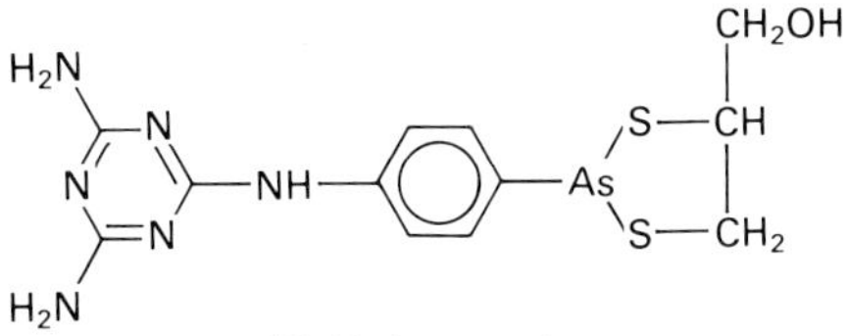

(**3**) Melarsoprol

The situation is even worse for South American trypanosomiasis (Chagas' disease): as we will see later there are no curative drugs. This disease occurs in the American continent between the latitudes of 42°N and 43°S: it is estimated that over 35 million people are at risk with at least 20 million infected. The disease and its treatment has been reviewed recently by Brener (1979, 1982), Ribiero-dos-Santos, Rassi and Koberle (1981), Brown (1983) and Kierszenbaum (1984). The reduviid bug vectors live in crevices, for example, in cracks in the walls or in the roofing material of the dwellings; the poorer the dwelling the greater the number of potential hiding places. The bugs sally forth at night to engorge with a blood meal, the parasite entering via a skin abrasion of the host, as indicated earlier. In common with the African trypanosome it is a 'metacyclic' morphological form which enters the host, rapidly transforming to the trypomastigote form which is elongated and has a flagellum and so is motile. Trypanosomes can be generally regarded as bloodstream parasites but some, such as *Trypanosoma cruzi* — the causative organism of Chagas' disease — have adapted to living within host cells. The trypomastigotes thus rapidly invade muscle tissue and the reticuloendothelial system. Here they convert to the amastigote morphological form, which is rounded in shape and has no flagellum. The amastigote is the replicative form; some invade surrounding cells, others revert back to the trypomastigote form and enter the bloodstream — these may invade other tissues or may be taken up by the insect vector as it takes a blood meal, so leading to the completion of the life cycle. The muscle tissue invaded varies with the strain of the organism; some strains invade heart muscle. This disease has an acute phase of about 1–2 months followed by a long chronic phase, the first part of which involves an apparent dormancy leading to a gross enlargement of the affected organ (probably due to an autoimmune effect), such as a ‘chagasic heart’, ‘chagasic oesophagus’ or ‘megacolon’.

There are three ways that can be used to control the disease: first, improved housing and spraying of dwellings with insecticide such as γ-BHC; second, eradication of the organisms in blood for transfusion (a possible mode of transmission of the disease); and, third, drug treatment. There is currently no prophylactic drug and chemotherapy is always going to be difficult to achieve as the replicative amastigote form is inaccessible in tissue. Only two drugs are

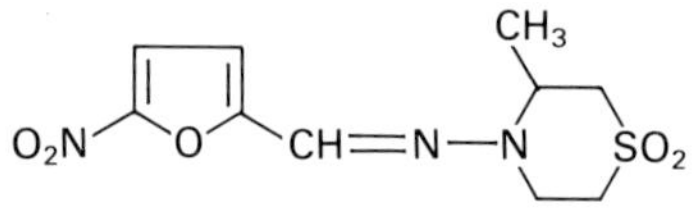

**(4)** Nifurtimox

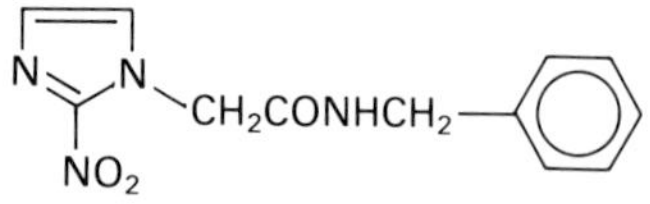

**(5)** Benznidazole

currently available, nifurtimox (**4**) introduced in 1976 and benznidazole (**5**) introduced in 1978. They are only active against trypomastigotes (i.e. the bloodstream form) and so are ineffective in the chronic phase, only being of any value in the acute phase. Treatment needs to be for 60 days and there are toxicity problems: peripheral neuropathy and central excitation is often seen with nifurtimox and benznidazole has an associated hypersensitivity manifested as dermatitis as well as neural effects and leucopenia. It is clear from consideration of the life cycle that drugs which are only effective against the trypomastigote form are inevitably not going to be fully effective. Once again new drugs are urgently

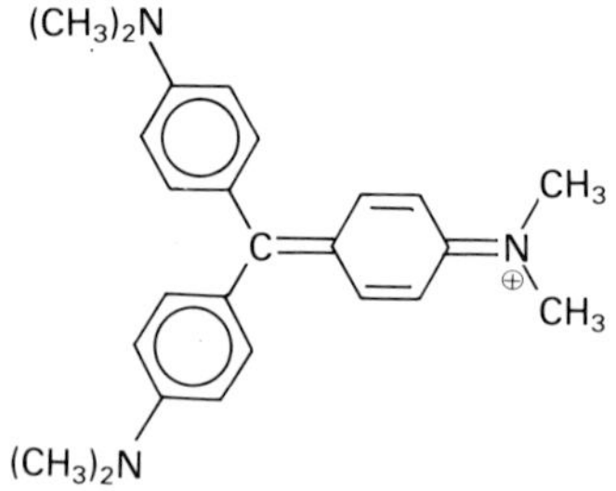

**(6)** Gentian violet

needed. Gentian violet (**6**) is commonly used to treat blood for transfusion and is effective. It does, however, have an effect on erythrocytes but has not been subjected to safety evaluation and can cause patients anxiety since it colours the blood purple. However, it could be replaced and Hammond and coworkers (1984), having screened over 500 drugs presently in clinical use for a variety of conditions, have found several groups of amphiphilic cationic drugs worthy of further investigation as possible sterilants for blood transfusion.

## 4.3 Leishmaniasis

The leishmaniases are a group of tropical diseases second only to malaria in importance. Like trypanosomal diseases, they are zoonoses and are transmitted by an insect vector, the sandfly. There are several different species of the genus *Leishmania* pathogenic to man, and together they give leishmaniasis a widespread prevalence such that it occurs in countries in the Old, the New and the Third World. There are about 400 000 new cases per annum. The leishmaniases and their treatment have been reviewed recently by D'Arcy and Harron (1983), Lainson (1983), Molyneux and Ashford (1983) and Marr (1984). The organism is in the promastigote morphological form in the proboscis of the sandfly and, on discharge into the host's blood during a blood meal, the resultant activation of complement attracts macrophages which phagocytose the protozoan organism which then divides as the amastigote morphological form within the macrophage. This eventually leads to release of amastigotes via rupture of the macrophage and parasitization of further macrophages. Sandflies are then infected by feeding off an infected person, leading to the completion of the cycle.

There are three major forms for the disease. Cutaneous leishmaniasis (synonyms Bagdhad boil, Delhi button, oriental sore) is caused by different species and subspecies in different geographical locations: *L. tropica* in Asia and the Mediterranean coast; *L. tropica aethiopica* in Ethiopia and in Kenya; *L.t. major* in Asia and North Africa; *L. mexicana mexicana*, *L.m. amazonensis*, *L.m. venezuelensis*, *L. braziliensis braziliensis*, *L.b. guyanensis*, *L.b. panamensis* and *L.b. peruviana* in Central and South America. The disease manifests as a skin lesion that might be self-healing, may ulcerate or in *L. braziliensis* subspecies may develop as large diffuse lesions. The second form, mucocutaneous leishmaniasis (espundia), is caused by *L.b. braziliensis* and starts as a skin lesion, invading the nasal and upper palate mucous membranes and spreading to cartilage and connective tissue giving severe facial mutilation through destruction of this tissue. The third form is visceral leishmaniasis (kala azar) caused by *L. donovani* and occurs in China, India, Southern Russia, the Mediterranean coast, East and Central Africa and South America. Spleen enlargement is the typical sign of this disease, which is severe and often fatal. Sometimes, e.g. where drug treatment is not completely successful, kala azar can manifest as a skin lesion (this is termed post kala azar dermal leishmaniasis). The only drugs effective against kala azar are the antimonial drugs sodium stibogluconate or meglumine antimonate. They are not ideal: they have to be given by intramuscular injection (i.m.) as a 10 day course and have severe side effects including cardiotoxicity and hepatotoxicity. Pentamidine is the second-line drug, with amphotericin (which is highly nephrotoxic) being used as a last-ditch treatment. These agents are relatively effective for visceral and generally also for cutaneous leishmaniasis (except the diffuse form) but less so for mucocutaneous leishmaniasis. Clearly there is again an urgent need for new drugs.

In the first three sections of this review it has been demonstrated that new drugs are urgently needed for the treatment of all forms of leishmaniasis and trypanosomiasis in man. An overview of the diseases has been given as it is necessary to have a comprehension of the diseases and the causative agents for a full appreciation of the rest of the review, where the potential for new drug design will be assessed by considering how the current drugs act and how parasite biochemistry differs from that of the host. As in many ways the organisms causing leishmaniasis and trypanosomiasis have similar biochemistries, they are considered together, rather than separately, in the rest of this review.

## 4.4 Inhibition of metabolic pathways

### *4.4.1 Carbohydrate metabolism*

Respiration in the bloodstream form of African trypanosomes (reviewed by Fairlamb, 1982) is insensitive to cyanide: this form of the organism contains no cytochromes and no haem. The single mitochondrion is defunct in this respect, until the organism is in the insect vector when the electron transport pathway develops (Vickerman, 1965). Since the organism is bathed in serum, it has a ready supply of glucose and so does not need maximal energy return. The Krebs cycle is incomplete, only some enzymes (e.g. malate dehydrogenase) being present (Falk, Akinrimisi and Onoagbe, 1980). By contrast, *T. cruzi* has a complete Krebs cycle and a fully functional mitochondrion throughout its life cycle (Rogerson and Gutteridge, 1980) with a high level of carbon dioxide fixation through carboxylation of phosphoenolpyruvate, so giving high succinate levels (Cazzulo *et al.*, 1980).

In African trypanosomes, glycolysis is therefore a crucial pathway (Fairlamb and Bowman, 1977; Fairlamb, Opperdoes and Borst, 1977) and makes an attractive target for selective attack, particularly since the rate of glycolysis is 50 times that in mammalian cells (Oduro, Bowman and Flynn, 1980). Both the rate-limiting step and differences between glycolysis in the parasite and in man are worthy of consideration: there are already data on both. It is the initial uptake of glucose that is the rate-limiting step (MacKenzie *et al.*, 1983) and there are two major novel features of the pathway in the parasite. First, the first nine enzymes, instead of being in the cytoplasm, are located in a microbody termed the glycosome (Opperdoes and Borst, 1977; Visser and Opperdoes, 1980) (Figure 4.1) which does not exclusively contain glycolytic enzymes. The glycosome appears to be a common feature of the Kinetoplastida under consideration since it is also found in *T. cruzi* and *Leishmania* species (Taylor *et al.*, 1980). The second difference between glycolysis in the parasite and man is that pyruvate is the end product — there is no lactate dehydrogenase. As the pyruvate → lactate step in man serves to regenerate $NAD^+$ (see Figure 4.1) an alternative mechanism is needed in the trypanosome — the so-called 'glycerophosphate shuttle'. Dihydroxyacetonephosphate is reduced to sn-glycerol-3-phosphate by the $NADH + H^+$, so regenerating $NAD^+$. Under aerobic conditions the sn-glycerol-3-phosphate is oxidized back to dihydroxyacetonephosphate by a glycerophosphate

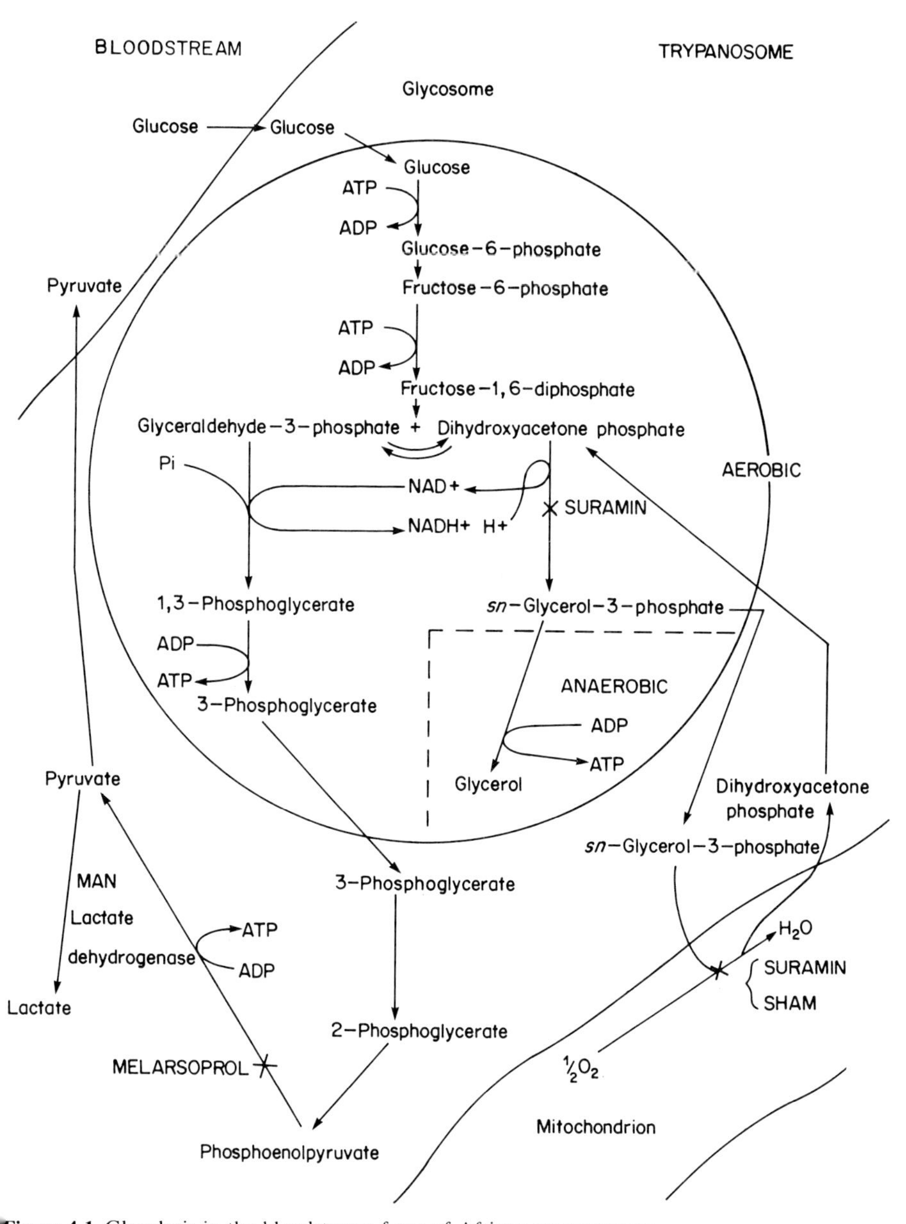


**Figure 4.1** Glycolysis in the bloodstream form of African trypanosomes.

oxidase system in the mitochondrion (Figure 4.1) and so it reenters the pathway. Under anaerobic conditions this cannot occur and instead a different pathway is used in which the sn-glycerol-3-phosphate is converted to glycerol, so generating ATP from ADP (Figure 4.1).

The assertion above that the dependency of the African trypanosomes on glycolysis makes it a target for drug action is borne out by the finding that suramin (**1**) is an inhibitor of both the glycerophosphate oxidase (Fairlamb and Bowman, 1977) and the glycerophosphate dehydrogenase (Fairlamb and Bowman, 1980a) (Figure 4.1). Also, melarsoprol is an inhibitor of the enzymes glycerophosphate oxidase, glycerol kinase and pyruvate kinase. It is thought to act in the cell primarily against the latter, the glycosome membrane presumably acting as a barrier to drug access to the former two enzymes (Fairlamb, 1982). Salicylhydro-

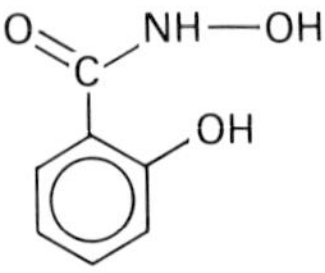

(**7**) Salicylhydroxamic acid (SHAM)

xamic acid (SHAM) (**7**) is another inhibitor of the glycerolphosphate oxidase (Figure 4.1), possibly by a metal chelation effect. However, administration of SHAM leads to a switch to the anaerobic pathway which can be blocked by coadministration of glycerol; the SHAM–glycerol combination thus being trypanocidal (van der Meer *et al.*, 1979). However, the amount of glycerol required (e.g. up to 1 litre i.v. in a goat) (van der Meer *et al.*, 1980) makes this impracticable. It can be appreciated that simultaneous blocking of the aerobic and anaerobic modes of metabolism of sn-glycerol-3-phosphate would be expected to be very effective in blocking reoxidation of NADH produced by oxidation of glyceraldehyde-3-phosphate. Disruption of glycolysis is readily detected experimentally as a reduction in the motility of the organism. Other leads include the findings that miconazole is an inhibitor of the glycerolphosphate oxidase, though only *in vitro* (Opperdoes 1980), and that cyclized thiourea derivatives of naphthoquinone are trypanocidal (Ulrich and Cerami, 1982). All in all there seems no reason why selective inhibitors of trypanosomal glycolysis cannot be developed.

### *4.4.2 Purine metabolism*

Trypanosomal and leishmanial organisms, indeed all parasitic protozoa so far examined, lack the pathway of synthesis of inosine monophosphate and so depend on obtaining purines from the host (reviewed by Hammond and Gutteridge, 1982; Marr, 1983; Marr and Berens, 1983; Ullman, 1984). A wide range of exogenous purines, purine nucleosides and purine nucleotides can be used (Fish *et al.*, 1982), the uptake process being saturable, and probably active, with a multiplicity of carriers (Hansen *et al.*, 1982; Okochi, Abaelu and Akinrimisi, 1983). The leishmanial cell surface at least has nucleotidase activity (Gottlieb and Dwyer, 1983). Hypoxanthine occupies a pivotal position in purine metabolism; a key pathway is the direct conversion of hypoxanthine to inosine monophosphate

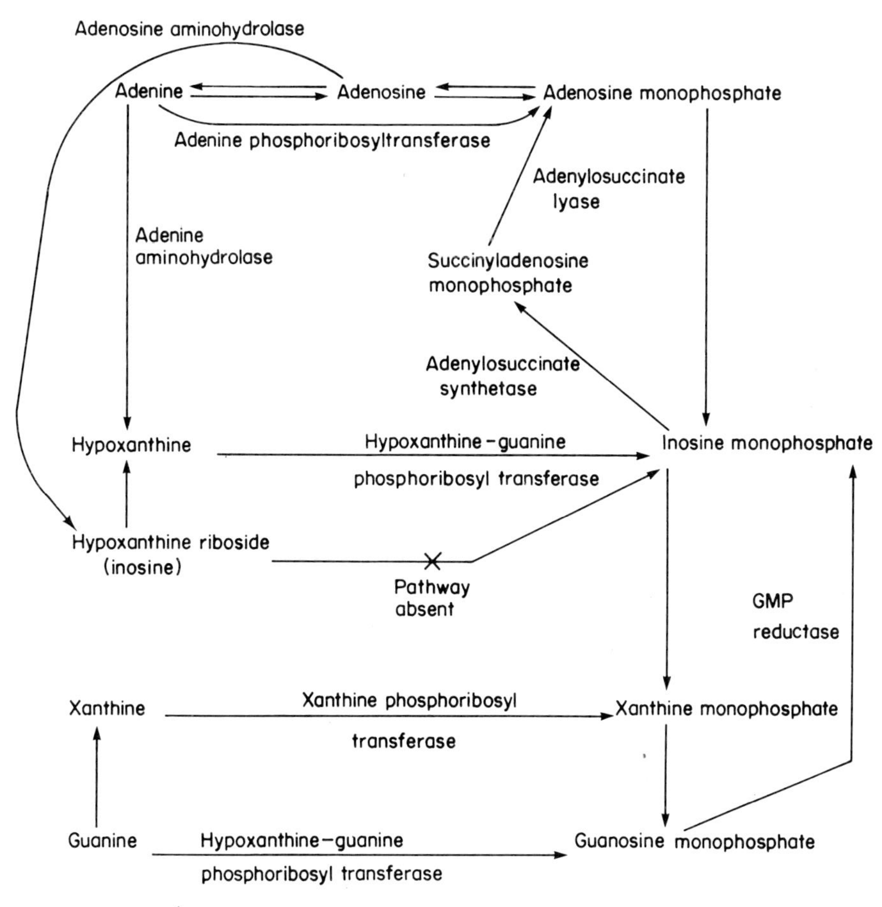


**Figure 4.2** Purine base, nucleoside and nucleotide interconversion pathways (generalized) in trypanosomes and leishmania.

(IMP) by hypoxanthine-guanine phosphoribosyl transferase (HGPRT). The general interconversions are indicated in Figure 4.2. There is a major difference between the pathways in trypanosomes and leishmania — the former, but not the latter, contain adenosine aminohydrolase whereas the reverse applies for adenine aminohydrolase (Davies, Ross and Gutteridge, 1983). Generalization is not fully valid, however; whilst the absence of *de novo* purine synthesis is a characteristic of all morphological forms, the salvage pathways may differ at the different stages of the life cycle (Hansen *et al.*, 1984) or between different species (Ogbunude and Ikediobi, 1983). We will not dwell on the differences since it is difficult to conceive how an agent could cause a total block of purine salvage. Nevertheless, an unsuccessful attempt has been made to prepare hypoxanthine derivatives as inhibitors of HGPRT (Piper *et al.*, 1980), any activity in the compounds prepared probably being via an effect on the membrane transport system. $N^6$-Methylade-

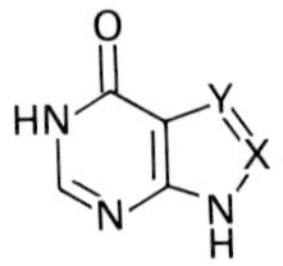

(**8a**) Allopurinol, X = N, Y = CH
(**8b**) Hypoxanthine, X = CH, Y = N

nine is a selective inhibitor of guanine deamination in protozoa (Nolan and Kidder, 1980).

More encouraging for drug development is the finding that the enzymes of purine interconversions in protozoa have different substrate specificity than those in man. This became apparent through work on allopurinol (**8a**), an isomer of hypoxanthine (**8b**). In man, allopurinol is rapidly cleared; 60 per cent of a dose is metabolized to oxipurinol by xanthine oxidase (the enzyme inhibited by allopurinol and oxipurinol), 30 per cent is excreted and 10 per cent is converted to its riboside, the latter not being further metabolized. Parasitic protozoa do not, however, have xanthine oxidase and allopurinol is converted to its riboside and its ribotide which then acts as an analogue of inosine monophosphate, being aminated, then further phosphorylated to the ATP analogue and incorporated into RNA (Berens, Marr and Brun, 1980; Avila, Avila and Argella de Casanova, 1981; Marr, 1983). Guanosine monophosphate (GMP) deamination is also inhibited (Spector *et al.*, 1984). The action on the protozoa is selective as macrophages infected with *L. tropica* were unaffected by allopurinol at concentrations which are antileishmanial (Berman and Webster, 1982).

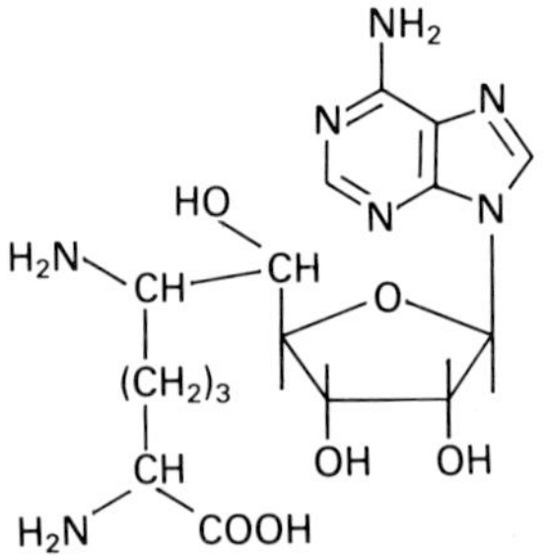

(**9**) Sinefungin

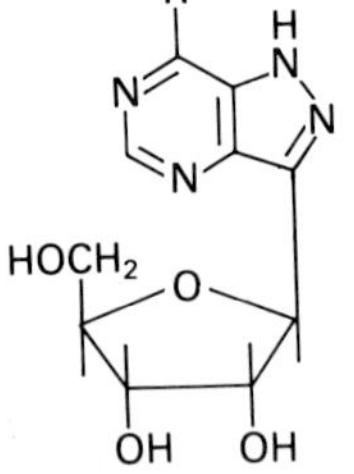

(**10a**) Formycin A, R = $NH_2$
(**10b**) Formycin B, R = OH

Other purines and purine analogues are also antiprotozoal (reviewed by Meshnick, 1984b), including thiopurinol (Spector and Jones, 1982), sinefungin (**9**) (Dube *et al.*, 1983), formycin A (**10a**) and formycin B (**10b**) (Berman *et al.*, 1983; Rainey, Garrett and Santi, 1983; Marr *et al.*, 1984). Formycin B does not just appear to give formycin A, as expected, but may also act by depletion of inosine monophosphate (Robinson *et al.*, 1984), though its ultimate effect is still on protein synthesis (Nolan, Berman and Giri, 1984).

Allopurinol is orally active in experimental systems; for example it gave healing in four out of five and cures in two out of five infections with *L.b. panamensis* in monkeys (Walton, Harper and Neal, 1983). However, whilst 14 cures were obtained in 17 cases of kala azar treated in North Bihar in India (Jha, 1983), when evaluated clinically against *L.b. braziliensis* in man it was disappointing (Marsden, Cuba and Barreto, 1984). This is not really surprising in view of its rapid clearance and the, self-defeating, increase of host hypoxanthine due to xanthine oxidase inhibition. What is required is a purine analogue (which has no effect on xanthine oxidase in the host), which itself is inactive but is activated selectively to a toxic species by the enzymes of the parasite purine salvage pathways. Alternatively, the lower substrate specificity could be exploited to design selective inhibitors of parasite enzymes which utilize the purine nucleotide monophosphates.

A brief mention should also be made of pyrimidine biosynthesis. Although these are synthesized *de novo* by parasitic protozoa (Hammond and Gutteridge, 1982) there is a difference in that some enzymes are in the glycosome. Also it is not inconceivable that the active site will have altered slightly during evolution; the enzymes may even be structurally different (Pascal *et al.*, 1983).

### *4.4.3 DNA-dependent processes*

Many of the established antiprotozoal agents are known to bind to DNA. There are two sites for binding in members of the Kinetoplastida: nuclear and kinetoplast DNA (reviewed by Williamson, 1979). As many anticancer agents react with DNA it is no surprise that anticancer agents are generally active against protozoa (Williamson and Scott-Finnigan, 1978; Kinnamon, Steck and Rane, 1979, 1980). Compounds which intercalate give DNA condensation whereas agents binding externally to the DNA lead to a 'sliced sausage' appearance to the DNA (Williamson, 1979). Some agents affect the kinetoplast; for example treatment with ethidium gives dyskinetoplastic forms (Riou, Belnat and Bernard, 1980). These are still viable since the kinetoplast DNA codes for cytochromes required during the part of the life cycle in the insect vector. Photoaffinity labelling studies suggest that acridines also affect the kinetoplast (Firth *et al.*, 1984). Other agents affect both the nucleus and the kinetoplast, e.g. diamidines such as pentamidine (**2**) (Williamson and McLaren, 1978) which are external binding agents and a group of bisquinolines which cause DNA clumping (Berman, Oka and Aikawa, 1984). Hydroxyurea, an inhibitor of reduction of ribonucleotide diphosphates to deoxyribonucleotide diphosphates, is interesting in that it only affects the nucleus but not the kinetoplast. Cells are produced with large nuclei and multiple kinetoplasts (Brun, 1980). Metal-containing complexes such as *cis*-platin (**11**) (Wysor *et al.*, 1982) and other experimental metal complexes (Farrell, Williamson and McLaren, 1984) also have antiprotozoal

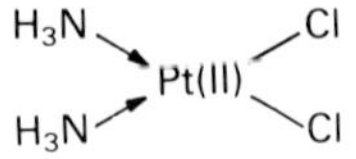

(**11**) *cis*-Platin

activity. In general, then, DNA-binding agents would be expected to be active against protozoa but toxicity is a major factor to consider.

It is assumed that binding to DNA leads to inhibition of DNA-dependent processes directly. However this may be an oversimplification, it is generally accepted now that intercalating agents induce topisomerase II-mediated strand breaks in DNA. This may provide fertile ground for drug research. Already,

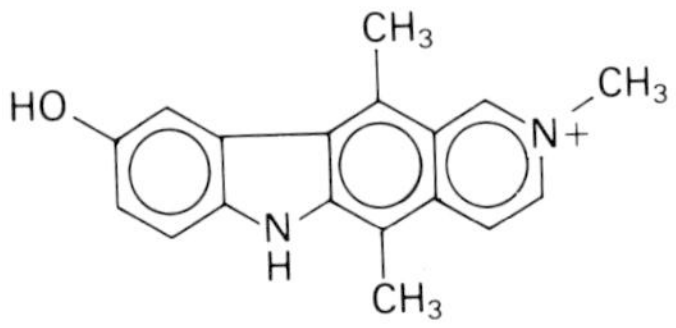

(**12**) Dimethylhydroxyellipticinium

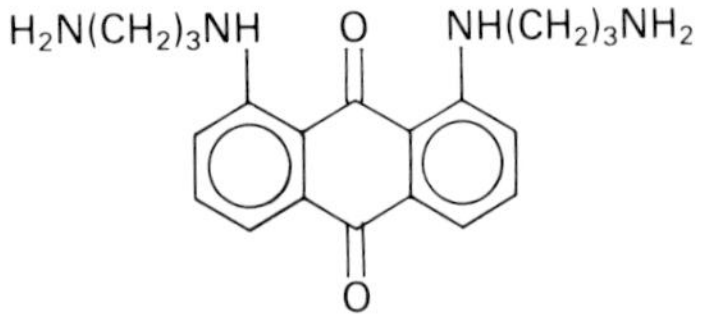

(**13**) Aminoalkylaminoanthroquinone

dimethylhydroxyellipticinium (**12**) has been shown to be active against *T. cruzi* by inhibition of topoisomerase I, with some apparent selectivity (Douc-Rasy, Kayser and Riou, 1983). Agents binding externally to DNA often have an additional effect on polyamine metabolism, e.g. the diamidines and amicarbalide (Nathan *et al.*, 1979). Even potential intercalators such as the aminoalkylaminoanthraquinone (**13**), which has antileishmanial activity, could either be acting by binding to DNA or by disrupting polyamine metabolism (Schnur *et al.*, 1983). There is a need therefore to fully characterize the mechanism of antiprotozoal action of DNA-binding agents. What is clear, however, is that any agent that damages or inactivates DNA in protozoa will be lethal.

One aspect that has as yet received scant attention is inhibition of enzymes which utilize DNA. This promises to be a fertile area as alluded to above with reference to the topoisomerases. Considering polymerases, the DNA polymerase system in the parasite differs from that in the host (Chang *et al.*, 1980), as do the DNA-dependent RNA polymerases (Kitchin, Ryley and Gutteridge, 1984).

### *4.4.4 Ornithine decarboxylase*

The enzyme ornithine decarboxylase (ODC) is produced early in the process of cell proliferation and so presents a logical target for design of inhibitors. Workers at Merrell showed that α-methylornithine (**14b**) is an inhibitor of ODC and further development aimed at generating suicide inhibitors led to α-difluoromethylornithine (**14a**) (DFMO). Ornithine decarboxylase converts ornithine to putrescine and is the initial reaction in the pathway of formation of the polyamines (Figure

COOH
|
$H_2N—C—R$
|
$(CH_2)_3$
|
$NH_2$

(**14a**) α-Difluoromethylornithine, R = $CHF_2$
(**14b**) α-Methylornithine, R = $CH_3$
(**14c**) α-Hydrazinoornithine, R = $NHNH_2$

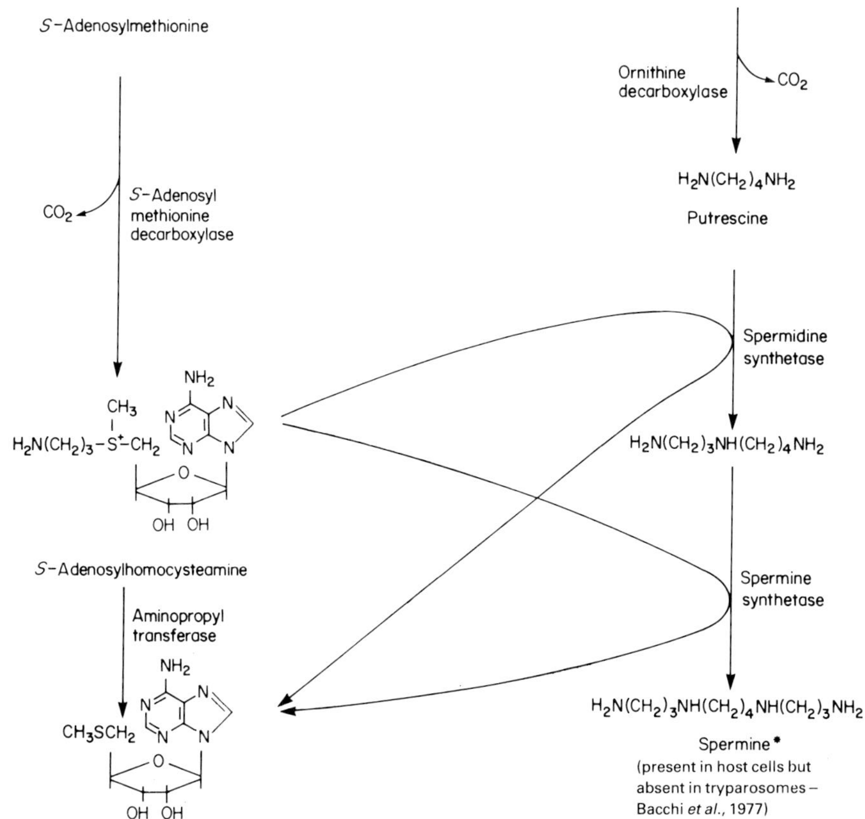


**Figure 4.3** Polyamine biosynthesis in trypanosomes.

4.3). ODC is regulated by the inhibitory protein antizyme but DFMO has no effect on the antizyme/ODC interaction. It acts directly on the enzyme (Kyriakidis *et al.* 1984). Decarboxylation of DFMO by the enzyme generates an electrophile which then alkylates a nucleophilic centre in the enzyme (Figure 4.4) (reviewed by McCann *et al.*, 1981; Fozard and Koch-Weser, 1982). That the action is due to the inhibition of polyamine synthesis is confirmed by the reversal of the inhibition by added polyamines (Nathan *et al.*, 1981).

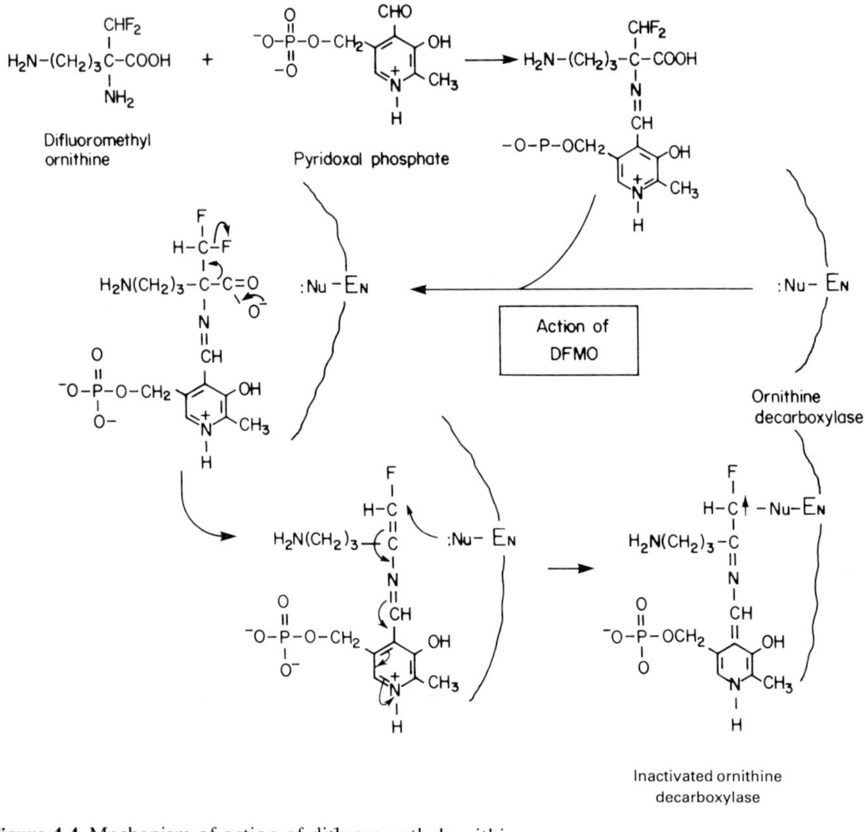

**Figure 4.4** Mechanism of action of difluoromethylornithine.

DFMO is particularly active against all the African trypanosomes (Karbe *et al.*, 1982) but has the disadvantage of having a very short half-life: in experimental animals it has been given in the drinking water at a 1 or 2 per cent concentration (Bacchi *et al.*, 1980). The effectiveness of DFMO against African trypanosomes is not only due to the high proliferation rate of the parasites but also to the fact that the enzyme in African trypanosomes is more sensitive than that in other organisms; for example ODCs from *E. coli*, *T. cruzi* and *L. donovani* promastigotes are insensitive to DFMO though that from *L. donovani* amastigotes is inhibited by DFMO (McCann *et al.*, 1981; Kallio, McCann and Bey,

1982; Coombs, Hart and Capaldo, 1983). DFMO is remarkably non-toxic, in some part due to the 60-fold lower inhibitory effect against the host enzyme than that of the trypanosome (Garofola *et al.*, 1982). It is, however, only trypanostatic. It promotes the long slender → short stumpy transformation (Bacchi *et al.*, 1983). The action can therefore be viewed as a disabling effect, the immune response acting to eliminate the parasitic organisms (de Gee, McCann and Mansfield, 1983). DFMO has now been evaluated in cases of sleeping sickness in man, with exciting results, and a lack of serious toxicity notwithstanding a daily dose of 10–15 g (Sjoerdsma and Schechter, 1984). Clearly, compounds with a longer half-life are needed. The most important aspect of the studies are that it appears to be active in man against organisms in the CNS (Sjoerdsma and Schechter, 1984): this is paralleled by activity against CNS-located trypanosomes in experimental animals when given in combination with other agents, e.g. bleomycin (Clarkson *et al.*, 1983).

Many analogues of DFMO have been studied (Mamont, Bey and Koch-Weser, 1980, Stevens and Stevens 1980, Casasa *et al.*, 1982) including α-methyl- (**14b**) and hydrazinoornithines (**14c**), α-ethynylputrescine (**15**) and dehydroputrescine (**16**). Oligoamino compounds such as amicarbalide and imidocarb (Bacchi *et al.*, 1981) and the other compounds indicated in the previous section (4.3) also appear to be inhibitors of polyamine utilization.

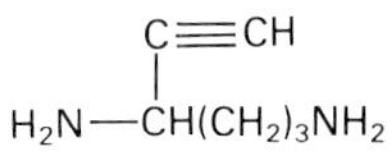

$$H_2N{-}CH(C{\equiv}CH)(CH_2)_3NH_2$$

(**15**) α-Ethynylputrescine

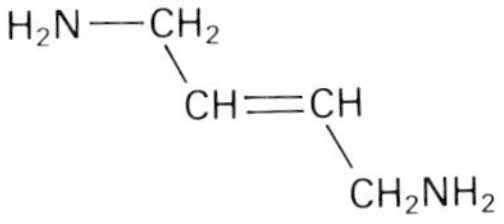

$$H_2N{-}CH_2{-}CH{=}CH{-}CH_2NH_2$$

(**16**) Dehydroputrescine

An alternative target is the *S*-adenosylmethionine decarboxylase enzyme (Figure 4.3). The antitumour compound methylglyoxalbisguanylhydrazone (Megag) is a known inhibitor (Stevens and Stevens, 1980). Similarly, the aminopropyltransferases (Figure 4.3) are a target (Pegg and Coward, 1981): transition state analogues provide one approach to inhibitor design (Pegg, Tang and Coward, 1982).

An attempt has been made to prepare inhibitors which are hybrids of pentamidine and Megag (Ulrich and Cerami, 1984). However, the best approach must be to achieve a simultaneous block of more than one enzyme in the pathway by agents which do not give a compensatory increase in ODC levels. The feedback repression must be maintained. The results to date on DFMO suggest that this is an important new agent against African trypanosomes.

### *4.4.5 Differentiation*

African trypanosomes are pleiomorphic, i.e. they exist in more than one form — the long slender and short stumpy forms. *Trypanosoma cruzi* exists in both the amastigote and trypomastigote form in man. We have seen above that DFMO can affect differentiation in African trypanosomes, promoting the transition to the short stumpy form. Differentiation is one aspect of protozoal metabolism that begs investigation. Evidence is scanty concerning the processes which trigger differentiation; possibly cyclic-AMP is a regulator of differentiation (Walter and Opperdoes, 1982). Prostaglandins possibly have an antidifferentiation role since indomethacin has been shown to promote the long slender → short stumpy transition (Jack *et al.*, 1984).

The enzyme ADP-ribosyl transferase (ADPRT-ase) appears to have a central role in differention by activating the relevant genes. Inhibitors of ADPRT-ase, such as 3-methoxybenzamide and 5-methylnicotinamide, have indeed been shown to block differentiation, but not proliferation, of *T. cruzi* amastigotes (Williams, 1984). The design of potent specific inhibitors of protozoal ADPRT- ase is worthy of consideration. This will become even more so when we consider the immune response later in this review (Section 4.6).

### 4.4.6 Other processes

Perhaps one of the most interesting recent findings is that tricyclic sedatives and antidepressants such as chlorpromazine (**17a**), trifluoperazine (**17b**), thioridazine (**17c**) and clomipramine (**18**) have antileishmanial activity (Pearson *et al.*, 1982,

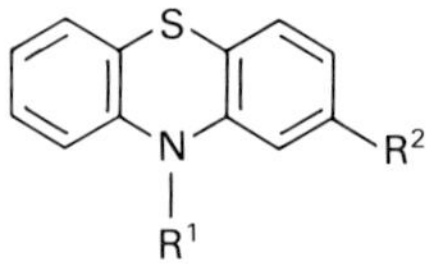

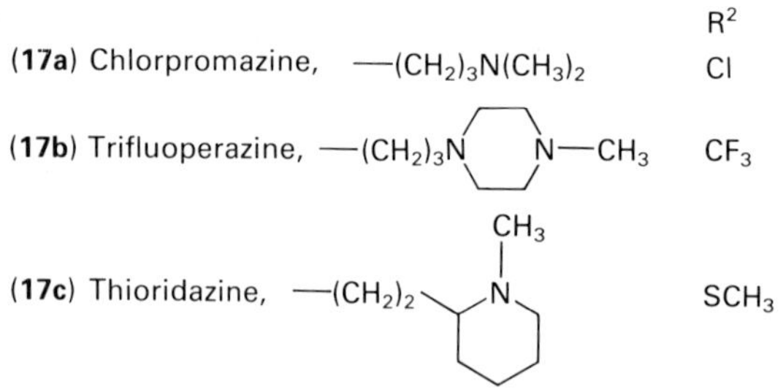

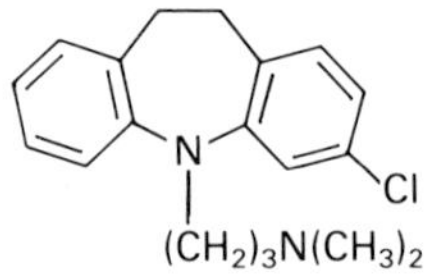

| | $R^1$ | $R^2$ |
|---|---|---|
| (**17a**) Chlorpromazine, | $—(CH_2)_3N(CH_3)_2$ | Cl |
| (**17b**) Trifluoperazine, | $—(CH_2)_3N$ (piperazine) $N—CH_3$ | $CF_3$ |
| (**17c**) Thioridazine, | $—(CH_2)_2$– (N-$CH_3$ piperidine) | $SCH_3$ |

(**18**) Clomipramine — N-$(CH_2)_3N(CH_3)_2$, Cl

1984; Zilberstein and Dwyer, 1984), being active not only against organisms in the bloodstream but also against amastigotes within macrophages. Due to the high dose (*circa* ten times the therapeutic dose) that would be required it is unlikely that these compounds will find an antileishmanial use (Pearson *et al.*, 1984) except perhaps topically (Henriksen and Lende, 1983). The work provides a lead for further drug development. The antiprotozoal effect has been suggested to be due to inhibition of the transmembrane proton pump which drives active transport systems (Zilberstein and Dwyer, 1984) or due to an effect on calmodulin (Ruben *et al.*, 1984). The latter suggestion comes from work with African trypanosomes which, interestingly, have been shown to have calmodulin which differs from that of the host (Ruben, Egwuaga and Patten, 1983). After treatment with phenothiazines, African trypanosomes rapidly lose their motility, the cells round up and there is disintegration of the pellicular microtubules (a system that runs under the plasma membrane) but not of the axonemic microtubules (Seebeck and Gehr, 1983). The effect therefore seems to be on the submembrane microtubule system: this is a cellular feature unique to the organisms.

Several other metabolic processes have also been identified in the parasitic protozoa which could provide possible points for selective attack. These include the use of threonine by African trypanosomes as an energy source (Klein *et al.*, 1980), a threonine dehydrogenase yielding acetyl CoA and glycine (Gilbert, Klein and Miller, 1983). Regarding *T. cruzi*, this has been shown to possess a unique anabolic NADP glutamate dehydrogenase and a unique catabolic NAD glutamate dehydrogenase (Cazzulo *et al.*, 1979). Gossypol has been shown to be an inhibitor of oxidoreductases such as these (Gerez de Burgos *et al.*, 1984). We do not necessarily need to focus on systems unique to the parasite. It is possible that the same enzyme in the host and the parasite have different active site structures or that different enzyme systems catalyse a step common to the host and parasite. An example of the latter is that in *Crithidia fasciculata*, and by inference other protozoa, dihydrofolate reductase and thymidylate synthetase activity are present in a single protein whereas different enzyme proteins are present in man (Ferone and Roland, 1980). Cofactors may also differ and so provide a lead for drug design; for example trypanosomes and leishmania contain a unique cofactor, trypanothione, for glutathione reductase (Fairlamb *et al.*, 1985).

The dependence of, for example, trypanosomes on exogenous compounds (fatty acids, cholesterol, purines, prophyrins) (Voorheis, 1980; Salzman *et al.*, 1982) provides a site for attack — namely the active transport systems. Cell surface proteases provide an additional site for consideration for drug design; two proteases (cysteine proteases) have been identified in amastigotes of *L.m. mexicana* with only one being present in promastigotes (Pupkis and Coombs, 1984), with a similar proteolytic system having been identified in *T. cruzi* (Bontempi *et al.*, 1984). Leupeptin and antipapain are inhibitors of one or more of these enzymes (Coombs and Baxter, 1984). Whilst considering the cell surface, it should be mentioned that a particular protozoal species only parasitizes a small range of hosts, and in the non-natural host the organism is lysed by serum high-density lipoprotein (Rifkin, 1983, 1984). If it were possible to mimic this effect then an effective antiprotozoal action would be obtained.

Several leads which are chemical rather than biochemical have been derived from testing known drugs. Ketoconazole (Weinrauch *et al.*, 1983, Raether and Seidenath, 1984), rifampicin (El-on *et al.*, 1983) and silver sulphonamides (Wysor and Scovill, 1982) have been shown to have activity against *L. major* and *L. donovani*, *Leishmania* species and *T.b. rhodesiense* respectively. Alternatively, synthetic programmes have generated compounds active against African

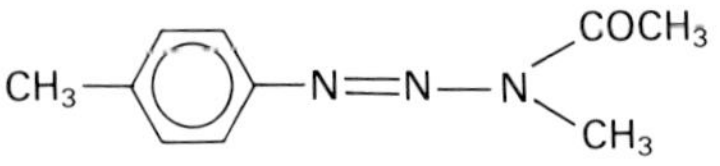

(**19**) Triazene

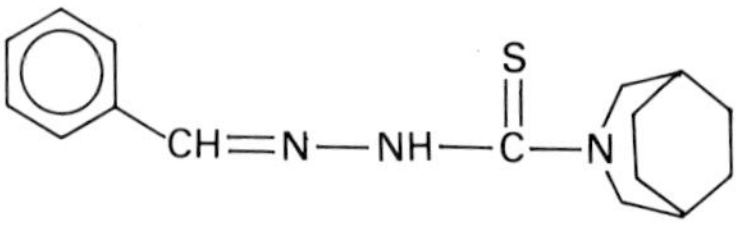

(**20**) Thiosemicarbazone

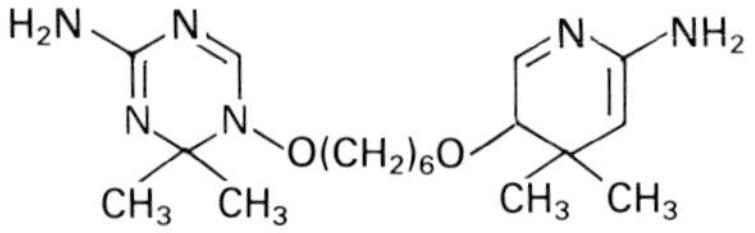

(**21**) Triazine

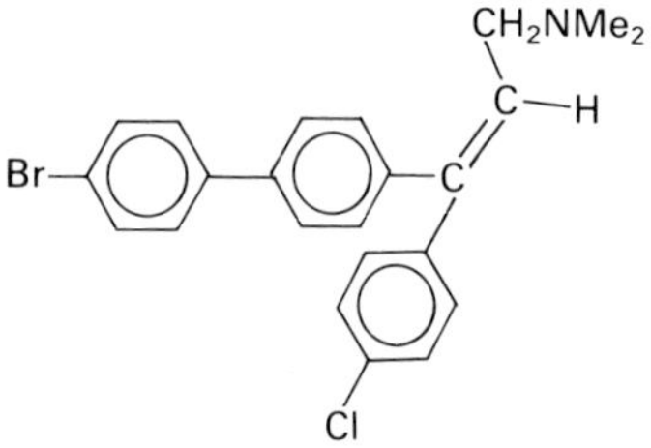

(**22**) Cinnamoylamine

trypanosomes, such as the triazene (**19**) (Dunn *et al.*, 1980), the thiosemicarbazone (**20**) (Casero *et al.*, 1980), the triazine (**21**) (Knight and Ponsford, 1982) and the cinnamoylamine (**22**) (Barrett *et al.*, 1982) active against *T. cruzi*. *Pseudomonas fluorescens* contamination of *T. cruzi* has been shown to give lysis and an antiprotozoal factor has been isolated (Mercado and Colon-Whitt, 1982).

A further problem is that occurring in late-stage African trypanosomiasis, i.e. the typical CNS 'sleeping sickness' effect. Identification of the mechanism could give a lead on symptomatic or even therapeutic treatments. Dopamine release or tryptophol production have been suggested as the causal factor (Stibbs, 1984a, 1984b). Tryptophol, and also indole acetic acid, are formed from tryptophan by an enzyme system differing from that of the host (Stibbs and Seed, 1975), presenting yet another target.

### 4.5 Oxidative damage

The possibility of inducing oxidative damage in kinetoplastids is of interest for two reasons. First, leishmanial organisms live in macrophages and so must avoid the oxidative burst consequent on phagocytosis by macrophages. (The cell-mediated killing of protozoa is reviewed by Thorne and Blackwell, 1983). Second, African trypanosomes have no catalase nor glutathione peroxidase so should be more sensitive to peroxide than are host cells.

Considering first the invasion of host cells by *T. cruzi* and *Leishmania* species, the initial stage is recognition by a lectin on the protozoan of *N*-acetylglucosamine (Crane and Dvorak, 1982; Bray, 1983a) or galactose (Villalta and Kierszenbaum, 1984) or mannose (Villalta and Kierszenbaum, 1983) on the host cell surface. The process may involve coating of the organism with fibronectin (Ouaissi *et al.*, 1984). Differences in the lectin-like substances on different strains of *T. cruzi* would explain why different strains give infections in different organs in the host (Araujo, Handman and Remington, 1980; Schottelius, 1982). Since *T.*

*cruzi* and leishmanial cells are rapidly killed in serum (Bray, 1983b), inhibition of the cell recognition process and the consequent delay in the parasite gaining sanctuary is an attractive strategy for protozoal cell kill.

The specific cell recognition process seems to be the mechanism by which triggering of the respiratory burst is avoided. Trypomastigotes of *T. cruzi* have binding sites differing from those on other morphological forms of the parasite (Katzin and Colli, 1983; Zenian and Kierszenbaum, 1983), i.e. different lectins are expressed by different morphological forms (Jackson and Diggs, 1984). Similarly, *L. donovani* promastigotes bind to different receptors on the macrophage from those for amastigotes (Channon, Roberts and Blackwell, 1984); only the amastigotes avoid stimulating the respiratory burst. It seems that the respiratory burst is not triggered rather than the host cell has a suppressed oxidative killing effect (Bray *et al.*, 1983; Docampo *et al.*, 1983). Hence if the oxidative burst could be triggered this would provide an attractive antiprotozoal mechanism. Several redox compounds such as methylene blue and phenazine methosulphate cause killing of intracellular leishmania or *T. cruzi* in experimental systems by stimulation of macrophage oxidative attack rather than a direct effect on the parasite (Rabinovitch *et al.*, 1982; Alves and Rabinovitch, 1983; Ryter *et al.*, 1983; Mauel, 1984; Mauel *et al.*, 1984). Once phagocytosed the two types of organism have different ways of avoiding subsequent possible enzymic attack when the parasitophorous vacuole fuses with lysosomes. *Trypanosoma cruzi* escapes into the cytoplasm before fusion can occur (Blackwell and Alexander, 1983) whereas leishmania presumably can resist lysosomal enzyme attack (Brazil, 1984).

African trypanosomes and *T. cruzi* contain no catalase and no glutathione peroxidase and hence are susceptible to oxidative and free radical attack. Indeed this is most probably the mode of action of nifurtimox. One electron reduction gives the nitro-anion radical ($RNO_2 \rightarrow RNO_2^{\dot{-}}$). This is reoxidized by oxygen which is thus reduced to superoxide ($O_2^{\dot{-}}$): superoxide dismutase then generates peroxide ($2O_2^{\dot{-}} + 2H^+ \rightarrow H_2O_2 + O_2$) leading to the hydroxyl radical ($H_2O_2 + O_2^{\dot{-}} \rightarrow OH^{\dot{-}} + OH^{\dot{-}} + O_2$) (Docampo and Stoppani, 1979; Docampo and Moreno, 1984a). Benznidazole, however, possibly does not act in this way (Docampo and Moreno, 1984b). Whilst there might be some further development possible with nitro compounds (Filardi and Brener, 1982), particularly in combination with, for example, suramin (**1**), for African trypanosomiasis (Jennings *et al.*, 1983), there is possibly greater scope for development with other, less thoroughly investigated compounds which are reductively activated by reduction and so generate $O_2^{\dot{-}}$. Alternatively, the known differences between protozoal and host superoxide dismutase (Docampo and Moreno, 1984) could be exploited. Considering other compounds, gentian violet is thought to act by generation of a free radical species (Docampo *et al.*, 1983) and a series of *p*-benzoquinoneimines has been synthesized, as peroxide generators, and shown to be trypanocidal (Grady *et al.*, 1984).

Finally, because of the lack of catalase and glutathione peroxidase, the parasitic organisms are dependent on glutathione for elimination of peroxide as well as for free radical and electrophilic species. The unique nature of the glutathione

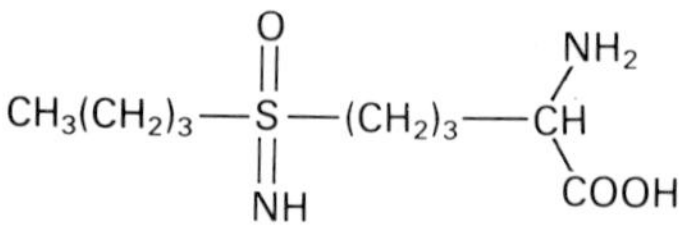

$$CH_3(CH_2)_3-\overset{\overset{\displaystyle O}{||}}{\underset{\underset{\displaystyle NH}{||}}{S}}-(CH_2)_3-CH(NH_2)COOH$$

(**23**) Buthionine sulphoxime

reductase cofactor was noted earlier (Section 4.4.6) Buthionine sulphoxime (**23**) is an inhibitor of glutamylcysteine synthetase, the first enzyme in glutathione synthesis: it appears to be selective for the trypanosomal enzyme (Arrick, Griffiths and Cerami, 1981). It is, however, a reversible inhibitor — a suicide inhibitor could well be a possible way of developing more potent compounds which capitalize on this lead.

In summary, current knowledge suggests there is potential for oxidative killing of protozoal parasites. Current agents are not satisfactory but provide good leads for further work.

## 4.6 The immune response

In order for the parasitic protozoa to enjoy the luxurious environment provided by the host, they must dodge the immune system. Two main techniques are used: either the organism lives in an intracellular site (*T. cruzi* and *Leishmania* species) or it is completely coated with a layer of variable surface glycoprotein (VSG — or variable antigen types, VAT), as in the case of African trypanosomes. There is also a 'smokescreen' mechanism whereby, for example, *T. cruzi* sheds surface antigen so neutralizing circulating antibody (Araujo, Chiari and Dias, 1981). Indeed it may be accumulation of antibody on host cells which leads to the autoimmune effect which is the pathogenic event. Additionally, 'suppressor' substances are released (Cunningham and Kuhn, 1980) which probably act by stimulating macrophage proliferation with a consequent reduction in the specific antibody response (Clayton *et al.*, 1980; Wellhausen and Mansfield, 1980; Grosskinsky *et al.*, 1983; Askonas, 1984).

Those organisms living intracellulary have to 'run the gauntlet' of the bloodstream where both humoral and cell-mediated responses lie in wait (Hatcher and Kuhn, 1982). In the case of *T. cruzi*, the acute phase terminates when circulating *T. cruzi* cells are eliminated, the chronic phase being characterized by multiplication of intracellular protozoa. This all indicates that, in time, development of vaccines against *T. cruzi* and leishmania should be feasible (Handman and Mitchell, 1985).

African trypanosomes survive in the bloodstream by keeping one step ahead of the immune response by varying the surface antigen. They have a large repertoire (>1000) of genes coding for VSGs, in fact constituting about 10 per cent of the total genome. Whilst at any one time the majority of organisms will be expressing the same single VSG, a small proportion will have switched to a new gene — a switch occurs in every $10^5$–$10^6$ cell divisions. Antibody developed against the first antigen will eliminate most of the parasites, those with the 'new' VSG proliferate, and so the antigen variation process continues. The antigen gene switching has been of considerable interest to molecular biologists and has been well reviewed (e.g. Cross, 1978; Turner, 1980, 1982, 1985; Bernards, 1982, 1984; Englund, Hajduk and Marini, 1982; Borst *et al.*, 1983; Donelson and Rice-Ficht, 1985). Gene duplication occurs, followed by transposition of this 'expression linked copy' to an expression site (DeLange and Borst, 1983; Milhausen *et al.*, 1983; Parsons *et al.*, 1983; Young *et al.*, 1983; Esser and Schoenbechler, 1985). Antigen switching is not a feature of *T. cruzi* (Snary, 1980). The process of antigen switching will not be considered in detail here, but the significance of this process for drug design needs to be considered. The VSG has a molecular weight of about 60 000 daltons: it is the *N* terminal region which varies. There is a hydrophobic 'handle' to anchor the VSG in the membrane and this is common to and invariant in all the VSGs. An inhibitor acting on this common hydrophobic anchor would therefore be of interest. The already known inhibitor of glycoprotein synthesis, tunicamycin (**24**), has been shown to inhibit mannose incorporation into VSGs but is only active at toxic levels (Casero, Porter and Bernacki, 1982).

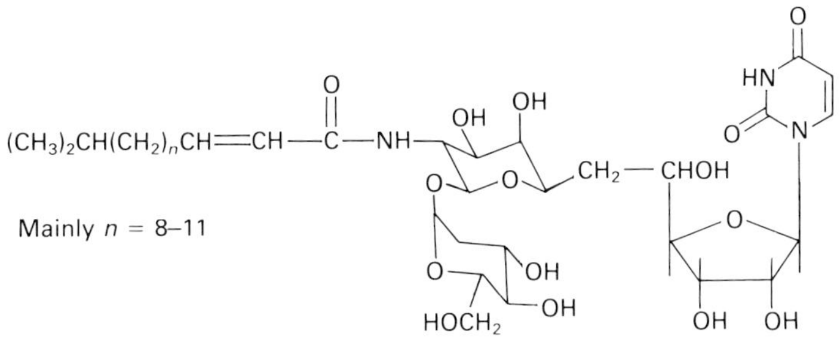

$(CH_3)_2CH(CH_2)_nCH{=}CH{-}C(=O){-}NH{-}$

Mainly $n$ = 8–11

(**24**) Tunicamycin

More ambitiously, the gene switching itself could be viewed as a target. The ADPRT-ase inhibitors discussed in Section 4.4.5 have the effect of reducing the frequency of switching (Farzaneh *et al.*, 1985). Finally, even with African trypanosomes, the possibility of vaccination is not totally out of the question. Cattle that have recovered from *T. brucei* infection are resistant to subsequent challenge and it is known that infective metacyclic trypanosomes only express a fraction of the total VSG repertoire (Barry, Crowe and Vickermann, 1983).

## 4.7 Selective delivery of drugs

In view of the fact that some protozoa, e.g. *Leishmania* species, infect professional phagocytes such as macrophages, it is surprising that little attention has been paid to the use of phagocytosis to get preferential uptake into the infected cells. The topic has been reviewed by Baillie (1984) and in general reviews on antiprotozoal drugs by Newton (1983) and Howells (1985). It has been shown that antimonial and other antileishmanial drugs encapsulated in liposomes have enhanced delivery to macrophages infected with leishmania (Alving, 1983). The fact that they are active against cutaneous as well as visceral leishmaniasis (New, Chance and Heath, 1981, 1983) suggests that drug-loaded macrophages can become infected so are then themselves acting as a drug-delivery system. An ultrastructural study with *L. donovani* (Weldon *et al.*, 1983) has shown that drug-loaded liposomes are endocytosed; the liposome-containing vesicle then fuses with the parasitophorous lyosome — so validating the 'delivery' mechanism. The delivery by the macrophage can be made more selective by incorporating moieties, such as mannose, for which the macrophage has surface receptors (Baillie, 1984).

Whilst this phagocytotic delivery is feasible for *Leishmania* species, for organisms such as *T. cruzi* there are considerable access problems. African trypanosomes, however, can be approached directly as they exist in the bloodstream. Liposomes have been shown to bind to, and be taken up by, African trypanosomes (Gruenberg *et al.*, 1979; Zumbutal and Weder, 1982); however, our own studies have been with macromolecules as drug carriers. These have the advantage of a greater stability and also drug can be covalently bound to the carrier. As seen in Section 4.4.6, there is a layer of pellicular microtubules underlying the membrane, but it is not present around the flagellar pocket, which acts as a rudimentary 'mouth'. Enzymes are excreted into the pocket which serves as the point of endocytosis of macromolecules (Langreth and Balber, 1975; Fairlamb and Bowman, 1980b; Steiger *et al.*, 1980).

As mentioned earlier (Section 4.4.3), there is a good correlation between anticancer and antiprotozoal activity; daunorubicin is an antitumour agent which is highly active *in vitro* against African trypanosomes but is inactive *in vivo* (Williamson and Scott-Finnigan, 1978). As the latter may be due to drug being taken up only transiently (Brown, Brown and Williamson, 1982), we coupled drug to protein in an attempt to simultaneously reduce plasma clearance and utilize the slower endocytotic uptake, mentioned above. The approach was successful: daunorubicin conjugated to protein is active *in vivo* but apparently only when the linkage is labile (Williamson *et al.*, 1981). Use of glutaraldehyde as a coupling agent gives active conjugates, whereas succinic anhydride then carbodiimide gives inactive conjugates. The former have chemically and enzymically labile, as well as stable, drug–protein linkages and can release free drug, whereas the latter having only stable amide linkages fail to release free drug (Hardman *et al.*, 1983a, 1983b).

Using an hplc assay (Brown, Wilkinson and Brown, 1981), the nucleus was shown to be the major site of drug location after administration of free (i.e. unconjugated) drug. After administration of a glutaraldehyde linked drug–ferritin conjugate, the majority of the conjugated drug (75 per cent) was found in the small particulate fraction whereas the daunorubicin released from the conjugate was mainly (80 per cent) located in the nucleus (Brown, Brown and Williamson, 1982; Brown *et al.*, 1982; Golightly *et al.*, 1983). Electron microscopy showed that nuclear effects typical of those seen with daunorubicin *in vitro*, namely nucleolar segregation and separation of peripheral chromatin from the nuclear membrane, are seen after *in vitro* and *in vivo* treatment with glutaraldehyde-linked drug–protein conjugates (Williamson, McLaren and Brown, 1983; Golightly *et al.*, 1985). The results are therefore fully consistent with a lysosomotropic mode of delivery, i.e. conjugate is endocytosed, drug being released in the lysosome and then exerting its effect on the nucleus. This use of labile drug–macromolecule conjugates is worth exploring further. Already Meshnick, Brown and Smith (1984) have shown that a similar effect can be achieved for *cis*-platin (**11**) — this drug being inactive against *Trypanosoma congolense* (an African trypanosome) as free drug, but active when coupled to polyglutamic acid by a labile linkage.

It can be seen that the selective delivery of drugs to parasitic protozoa is a field which is largely unexplored: indeed absorption and disposition as a whole has generally been ignored. Some pharmacokinetic work is now underway (Gilbert and Newton, 1982; Gilbert, 1983). It should be possible to improve the usage of current agents, based on the findings of such studies, or to achieve better drug release. As many antiprotozoal drugs are basic, ionic complexes can serve as depot preparations; also an isometamidium–dextran complex gives slow release of drug and less local effect at the injection site, in experimental animals (Aliu and Chineme, 1980).

## 4.8 Conclusions

Whilst this review has not attempted to be comprehensive, it has shown that the lack of new drugs for the treatment of trypanosomiasis and leishmaniasis is *not* due to the lack of possibility for selective attack: rather we are spoilt for choice when looking for possible targets. One part of the question posed in the Introduction has thus been answered. Considering the other parts of the question, it is also apparent that research into these diseases has not been misdirected, so the conclusion must be that there just has not been enough investment in this area. Since the diseases are Third World diseases, they are not prominent in the minds of drug developers: investment by drug companies would in any case be unwise. The financial return from drug sales would not cover the investment and, even if it did, there is no guarantee that Third World governments would choose to use some of their health budget in this way. Hence we are left with drugs which are more appropriate as material for students of the history of drug development than to therapy in the 1980s, and the need to finance the drug research by government, charitable or supranational bodies. New thinking is required — this is not a new realization since the current situation has been well reviewed (Cohen, 1979;

Goodwin, 1980; Steck, 1981; Fairlamb, 1982; Wang, 1982, 1984; Newton, 1983; Meshnick, 1984).

The research could well be carried out in academic and research institutes or alternatively/additionally by contracting the pharmaceutical industry to do some of the work. However, resources are going to be meagre and strategies must be tailored accordingly. Initial screening costs can be minimized by using *in vitro* screens now that *in vitro* culture methods are available (Hirumi, Doyle and Hirumi, 1977; Tanner, 1980; Hirumi and Hirumi, 1984; Neal and Croft, 1984). The 'blunderbuss' mass screening approach is, however, not going to be sufficiently economic. Although Desjardins and coworkers (1980) have indicated that the Walter Reed Institute is routinely screening compounds against *T.b. rhodesiense* we should recall that over 300 000 compounds were screened for antimalarial activity to get a new compound — mefloquine — into clinical use: even then, it is yet another quinine analogue. Because of the extent of our knowledge of protozoal biochemistry, we should use this area as a testing ground for the new technologies of drug delivery and engineering of specific chemical molecules to inhibit particular known biochemical systems. Interdisciplinary teams are needed for this to be effective.

In summary, the need for new drugs for the treatment of leishmaniasis and trypanosomiasis is desperate — millions are currently at risk. Academic and research institutions could make a big impact by application of the state-of-the-art drug design and drug-delivery techniques. It is a sobering thought that trypanosomes were the model system first used by Ehrlich, yet nearly a century later we have no fully acceptable drugs to treat trypanosomal diseases. There appears to be no reason why this cannot be remedied.

## 4.9 References


1. M.O. Abolarin, D.A. Evans, D.G. Tovey, and W.E. Ormerod, (1983). *Br. Med. J.*, **285**, 1380–1382.
2. Y.O. Aliu, and C.N. Chineme, (1980). *Toxicol. Appl. Pharm.*, **53**, 196–203.
3. M.J.M. Alves, and M. Rabinovitch, (1983). *Infect. Immun.*, **39**, 435–438.
4. C.R. Alving, (1983). *Pharmac. Ther.*, **22**, 407–424.
5. F.I.C. Apted, (1980). *Pharmac. Ther.*, **11**, 391–413.
6. F.G. Araujo, E. Chiari, and J.C.P. Dias, (1981). *Lancet*, **1981**, 246–249.
7. F.G. Araujo, E. Handman, and J.S. Remington, (1980). *J. Protozool.*, **27**, 397–400.
8. B.A. Arrick, O.W. Griffiths, and A. Cerami, (1981). *J. Expl Med.*, **153**, 720–725.
9. B.A. Askonas, (1984). *Parasitology*, **88**, 633–638.
10. J.L. Avila, A. Avila, and M. Argella de Casanova, (1981). *Mol. Biochem. Parasitol.*, **4**, 265–271.
11. C.J. Bacchi, J. Garafola, D. Mockenhaupt, P.P. McCann, K.A. Dieckma, A.E. Pegg, H.C. Nathan, E. Mullaney, L. Chunosoff, A. Sjoerdsma, and S.H. Hutner, (1983). *Mol. Biochem. Parasitol.*, **7**, 209–225.
12. C.J. Bacchi, H.C. Nathan, S.H. Hutner, D.S. Duch, and C.A. Nichol, (1981). *Biochem. Parmacol.*, **30**, 883–886.
13. C.J. Bacchi, H.C. Nathan, S.H. Hutner, P.P. McCann, and A. Sjoersma, (1980). *Science*, **210**, 332–334.
14. A.J. Baillie, (1984). *Pharm. Int.*, **5**, 168–172.
15. P.A. Barrett, E. Beveridge, D. Bull, I.C. Caldwell, P.J. Islip, R.A. Neal, and N.C. Woods, (1982). *Experientia*, **38**, 338–339.

16. J.D. Barry, J.S. Crowe, and K. Vickermann, (1983). *Nature (Lond.)*, **306**, 699–701.
17. R.L. Berens, J.J. Marr, and R. Brun, (1980). *Mol. Biochem. Parasitol.*, **1**, 69–73.
18. J.D. Berman, M. Oka, and M. Aikawa, (1984). *J. Protozool.*, **31**, 184–186.
19. J.D. Berman, L. Shee, R.K. Robins, and G.R. Revankar, (1983). *Antimicrob. Agents Chemother.*, **24**, 233–236.
20. J.D. Berman, and H.K. Webster, (1982). *Antimicrob. Agents Chemother.*, **21**, 887–891.
21. A. Bernards, (1982). *Trends in Pharm. Sci.*, **1982**, 253–255.
22. A. Bernards, (1984). *Biochim. Biophys. Acta*, **824**, 1–15.
23. J.M. Blackwell, and J. Alexander, (1983). *Trans. R. Soc. Trop. Med. Hyg.*, **77**, 636–645.
24. E. Bontempi, B.M. Franke de Cazzulo, A.M. Ruiz, and J.J. Cazzulo (1984) *Comp. Biochem. Physiol.*, **77B**, 599–604.
25. P. Borst, A. Bernards, L.H.T. Van der Ploeg, P.A.M. Michels, A.Y.C. Liu, T. DeLange, and J.M. Kooter, (1983). *Eur. J. Biochem.*, **137**, 383–389.
26. R.S. Bray, (1983a). *J. Protozool.*, **30**, 314–322.
27. R.S. Bray, (1983b). *J. Protozool.*, **30**, 323–329.
28. R.S. Bray, B. Heikal, P.M. Kaye, and M.A. Bray, (1983). *Acta Trop.*, **40**, 29–38.
29. R.P. Brazil, (1984). *Ann. Trop. Med. Parasitol.*, **78**, 87–91.
30. Z. Brener, (1979). *Pharmac. Ther.*, **7**, 71–90.
31. Z. Brener, (1982). *Bull. WHO*, **60**, 463–473.
32. J.E. Brown, J.R. Brown, and J. Williamson, (1982). *J. Pharm. Pharmacol.*, **34**, 236–239.
33. J.E. Brown, L.H. Patterson, J. Williamson, and J.R. Brown, (1982). *J. Pharm. Pharmacol.*, **34**, 42P.
34. J.E. Brown, P.A. Wilkinson, and J.R. Brown, (1981). *J. Chromatog.*, **226**, 521–525.
35. J.R. Brown, (1983). *Pharm. Internat.*, **4**, 53–57.
36. J.R. Brun, (1980). *J. Protozool.*, **27**, 122–128.
37. P. Casasa, C. Danzin, B.W. Metcalf, and M.J. Jung, (1982). *J. Chem. Soc. Chem. Commun.*, **1982**, 1190–1192.
38. R.A. Casero, J.L. Klayman, G.E. Childs, J.P. Scovill, and R.E. Desjardins, (1980). *Antimicrob. Agents Chemother.*, **18**, 317–322.
39. R.A. Casero, C.W. Porter, and R.J. Bernacki, (1982). *Antimicrob. Agents Chemother.*, **22**, 1008–1011.
40. J.J. Cazzulo, B.M.F. de Cazzulo, A.I. Hiya, and E.L. Seguara, (1979). *Comp. Biochem. Physiol.*, **643**, 129–131.
41. J.J. Cazzulo, E. Valle, R. Docampo, and J.J.B. Cannata, (1980). *J. Gen. Microbiol.*, **117**, 271–274.
42. L.M.S. Chang, E. Cherianthundum, E.M. Mahoney, and A. Cerami, (1980). *Science*, **208**, 510–511.
43. J.Y. Channon, M.B. Roberts, and J.M. Blackwell, (1984). *Immunology*, **53**, 345–355.
44. A.B. Clarkson, C.J. Bacchi, G.H. Mellow, H.C. Nathan, P.P. McCann, and A. Sjoerdsma, (1983). *Proc. Natn. Acad. Sci. USA*, **80**, 5729–5733.
45. C.E. Clayton, M.E. Selkirk, C.A. Corsini, B.M. Ogilvie, and B.A. Askonas, (1980). *Infect. Immunity*, **28**, 824–831.
46. S.S. Cohen, (1979). *Science*, **205**, 964–971.
47. G.H. Coombs, and J. Baxter, (1984). *Ann. Trop. Med. Parasitol.*, **78**, 21–24.
48. G.H. Coombs, T. Hart, and J. Capaldo, (1983). *J. Antimicrob. Chemother.*, **11**, 151–162.
49. M. St. J. Crane, and J.A. Dvorak, (1982). *Mol. Biochem. Parasitol.*, **5**, 333–341.
50. G.A.M. Cross, (1978). *Proc. R. Soc. Lond.*, **B202**, 55–72.
51. D.S. Cunningham, and K.E. Kuhn, (1980). *J. Immunol.*, **80**, 2122–2129.
52. P.F. D'Arcy, and D.W.G. Harron, (1983). *Pharm. Int.*, **4**, 238–242.
53. M.J. Davies, A.M. Ross, and W.E. Gutteridge, (1983). *Parasitology*, **87**, 211–219.
54. A.W.L. de Gee, P.P. McCann, and J.J. Mansfield, (1983). *J. Parasitol.*, **69**, 818–822.
55. T. DeLange, and P. Borst, (1983). *Nature (Lond.)*, **299**, 451–453.
56. P. de Raadt, (1976). *Trans. R. Soc. Trop. Med. Hyg.*, **70**, 114–116.
57. R.E. Desjardins, R.A. Casero, G.P. Willet, G.E. Childs, and C.J. Canfield, (1980). *Expl Parasitol.*, **50**, 260–271.
58. R. Docampo, A.M. Casellas, E.D. Madeira, R.L. Cardoni, S.N.J. Moreno, and R.P. Mason, (1983). *FEBS Lett.*, **155**, 25–30.
59. R. Docampo, and S.N.J. Moreno, (1984a). In *Free Radicals in Biology* (Ed. W.A. Pryor), Vol. 1, pp. 243–288, Academic Press, New York.

60. R. Docampo, and S.N.J. Moreno, (1984b). In *Oxygen Radicals in Biology and Medicine* (Eds W. Bors, M. Saran and D. Tait), pp. 749–752, Walter de Gruyter, Berlin.
61. R. Docampo, S.N.J. Moreno, R.P.A. Muniz, F.S. Cruz, and R.P. Mason, (1983). *Science*, **220** 1292-1295.
62. R. Docampo, and A.O.M. Stoppani, (1979). *Arch. Biochem. Biophys.*, **179**, 317–321.
63. J.E. Donelson, and A.C. Rice-Ficht, (1985). *Microbiol. Rev.*, **49**, 107–125.
64. S. Douc-Rasy, A. Kayser, and G. Riou, (1983). *Biochem. Biophys. Res. Commun.*, **117**, 1–5.
65. D.K. Dube, G. Mpimbaza, A.C. Allinson, E. Lederer, and L. Rovis, (1983). *Am. J. Trop. Med. Hyg.*, **32**, 31–33.
66. W.J. Dunn, J. Powers, J.B. Kaddu, and A.R. Njogu, (1980). *J. Pharm. Sci.*, **69**, 1465–1466.
67. J. El-On, E. Pearlman, L.F. Schnur, and C.L. Greenblatt, (1983). *Israel J. Med. Sci.*, **19** 240–245.
68. P.T. Englund, S.L. Hajduk, and J.C. Marini, (1982). *Ann. Rev. Biochem.*, **51**, 695–726.
69. K.M. Esser, and M.J. Schoenbechler, (1985). *Science*, **229**, 190–193.
70. D.A. Evans, (1981). In *Antibiotics and Chemotherapy*, Vol. 30, *Antiparasitic Chemotherapy* (Ed. H. Schonfeld), pp. 272–287, Karger, Basel.
71. A.H. Fairlamb, (1982). *Trends in Biochem. Sci.* **1982**, 249–253.
72. A.H. Fairlamb, P. Blackburn, P. Ulrich, B.T. Chaik, and A. Cerami, (1985). *Science*, **227** 1485–1487.
73. A.H. Fairlamb, and I.B.R. Bowman, (1977). *Expl Parasitol.*, **43**, 353–361.
74. A.H. Fairlamb, and I.B.R. Bowman, (1980a). *Mol. Biochem. Parasitol.*, **1**, 315–333.
75. A.H. Fairlamb, and I.B.R. Bowman, (1980b). *Expl Parasitol.*, **49**, 366–380.
76. A.H. Fairlamb, F.R. Opperdoes, and P. Borst, (1977). *Nature (Lond.)*, **265**, 270–271.
77. E. Falk, E.U. Akinrimisi, and I. Onoagbe, (1980). *Int. J. Biochem.*, **12**, 647–650.
78. N.P. Farrell, J. Williamson, and D.J. McLaren, (1984). *Biochem. Pharmacol.*, **33**, 961–971.
79. F. Farzaneh, S. Shall, P. Michels, and P. Borst, (1985). *Mol. Biochem. Parasitol.*, **14**, 251–259.
80. R. Ferone, and S. Roland, (1980). *Proc. Natn. Acad. Sci. USA*, **77**, 5802–5806.
81. L.S. Filardi, and Z. Brener, (1982). *Ann. Trop. Med. Parasitol.*, **76**, 293–297.
82. W.J. Firth, A. Messa, A. Reid, R.C. Wang, C.L. Watkins, and L.W. Yielding, (1984). *J. Med. Chem.*, **27**, 865–870.
83. W.R. Fish, D.L. Looker, J.J. Marr, and R.L. Berens, (1982). *Biochim. Biophys. Acta*, **719**, 223–231.
84. J.R. Foulkes, (1981). *Br. Med. J.*, **283**, 1172–1174.
85. J.R. Fozard, and J. Koch-Weser, (1982). *Trends in Pharm. Sci.*, **1982**, 107–110.
86. J. Garafola, C.J. Bacchi, S.D. McLaughlin, D. Mockenhaupt, G. Tineba, and S.H. Hutner, (1982). *J. Protozool.*, **29**, 389–394.
87. N.M. Gerez de Burgos, C. Burgos, E.E. Montamat, L.E. Povai, and A. Blanco, (1984). *Biochem. Pharmacol.*, **33**, 955–959.
88. R.J. Gilbert, (1983). *Br. J. Pharmacol.*, **80**, 133–319.
89. R.J. Gilbert, R.A. Klein, and P.G.G. Miller, (1983). *Comp. Biochem. Physiol.*, **74B**, 277–281.
90. R.J. Gilbert, and B.A. Newton, (1982). *Parsitology*, **85**, 127–148.
91. L. Golightly, J.E. Brown, J.B. Mitchell, L.H. Patterson, and J.R. Brown, (1983). *J. Pharm. Pharmacol.*, **35**, 33P.
92. L. Golightly, J.B. Mitchell, J.E. Brown, and J.R. Brown, (1985). *J. Pharm. Pharmacol.*, **37**, 145P.
93. L.G. Goodwin, (1980). *Trans R. Soc. Med. Hyg.*, **74**, 1–7.
94. M. Gottlieb, and D.M. Dwyer, (1983). *Mol. Biochem. Parasitol.*, **7**, 303–317.
95. R.W. Grady, S.H. Blobstein, S.R. Meshnick, P.C. Ulrich, A. Cerami, J. Amirmoazzani, and E.M. Hodnett, (1984). *J. Cell. Biochem.*, **25**, 15–29.
96. C.M. Grosskinsky, R.A.B. Ezekowitz, G. Berton, S. Gordon, and B.A. Askonas, (1983). *Infect. Immun.*, **37**, 1080–1086.
97. J. Gruenberg, D. Coral, A.L. Knupper, and J. Deschusses, (1979). *Biochem. Biophys. Res. Commun.*, **88**, 1173–1179.
98. D.J. Hammond, B. Cover, and W.E. Gutteridge, (1984). *Trans. R. Soc. Trop. Med. Hyg.*, **78**, 91–95.
99. D.J. Hammond, and W.E. Gutteridge, (1982). *Biochem. Biophys. Acta*, **718**, 1–10.
100. E. Handman, and G.F. Mitchell, (1985). *Proc. Natl Acad. Sci. US*, **82**, 5910–5914.
101. B.D. Hansen, J. Perez-Arbelo, T.F. Walkony, and L.D. Hendricks, (1982). *Parasitology*, **85**, 271–282.

102. B.D. Hansen, H.K. Webster, L.D. Hendricks, and M.G. Pappass, (1984). *Expl Parasitol.*, **58**, 101–109.
103. M.A. Hardman, L.H. Patterson, J. Williamson, and J.R. Brown, (1983a). *Biochem. Trans.*, **11**, 182.
104. M.A. Hardman, L.H. Patterson, J. Williamson, and J.R. Brown, (1983b). Second International Conference on Drug Adsorption, Edinburgh.
105. F.M. Hatcher, and R.E. Kuhn, (1982). *Science*, **218**, 295–296.
106. T-H. Henriksen, and S. Lende, (1983). *Lancet*, **i**, 126.
107. H. Hirumi, J.J. Doyle, and K. Hirumi, (1977). *Science*, **196**, 992–994.
108. H. Hirumi, and K. Hirumi, (1984). *Ann. Trop. Med. Parasitol.*, **78**, 327–330.
109. R.E. Howells, (1985). *Parasitology*, **90**, 687–703.
110. R.M. Jack, S.J. Black, S.L. Reed, and S.E. Davis, (1984). *Infect. Immun.*, **43**, 445–448.
111. P.R. Jackson, and C.L. Diggs, (1984). *J. Protozool.*, **30**, 662–668.
112. F.W. Jennings, G.M. Urquhart, P.K. Murray, and B.M. Miller, (1983). *Trans. R. Soc. Trop. Med. Hyg.*, **77**, 693–698.
113. T.K. Jha, (1983). *Trans. R. Soc. Trop. Med. Hyg.*, **77**, 204–207.
114. A. Kallio, P.P. McCann, and P. Bey, (1982). *Biochem. J.*, **204**, 771–775.
115. E. Karbe, M. Bottger, P.P. McCann, A. Sjoerdsma, and E.K. Frietas, (1982). *Tropenmed. Parasitol.*, **33**, 161–162.
116. A.M. Katzin, and W. Colli, (1983). *Biochim. Biophys. Acta*, **727**, 403–411.
117. F. Kierszenbaum, (1984). In *Parasitic Diseases*, Vol. 2, *The Chemotherapy* (Ed. J.M. Mansfield), pp. 133–163, Marcel Dekker, New York.
118. K.E. Kinnamon, E.A. Steck, and D.S. Rane, (1979). *Antimicrob. Agents Chemother.*, **15**, 157–160.
119. K.E. Kinnamon, E.A. Steck, and D.S. Rane, (1980). *J. Natn. Cancer Inst.*, **65**, 391–394.
120. P.A. Kitchin, J.F. Ryley, and W.E. Gutteridge, (1984). *Comp. Biochem. Physiol.*, **77B**, 223–231.
121. R.A. Klein, J.M. Angus, A.E. Amadife, and L. Smith, (1980). *Comp. Biochem. Physiol.*, **66B**, 143–146.
122. D.J. Knight, and R.J. Ponsford, (1982). *Ann. Trop. Med. Parasitol.*, **76**, 589–594.
123. D.A. Kyriakidis, F. Flamigni, J.W. Parolak, and E.S. Canellakis, (1984). *Biochem. Pharmacol.*, **33**, 1575–1578.
124. R. Lainson, (1983). *Trans. R. Soc. Trop. Med. Hyg.*, **77**, 569–596.
125. S.G. Langreth, and A.E. Balber, (1975). *J. Protozool.*, **22**, 40–53.
126. T.M. Leach, and C.J. Roberts, (1981). *Pharmacol. Ther.*, **13**, 91–147.
127. P.P. McCann, C.J. Bacchi, W.L. Hanson, G.D. Cain, H.C. Nathan, S.H. Hutner, and A. Sjoerdsma, (1981). In *Advances in Polyamine Research* (Eds. C.M. Calderera, V. Zappia and U. Bachrach), Vol. 3, pp. 97–110, Raven Press, New York.
128. N.E. MacKenzie, J.E. Hall, I.W. Flynn, and A.I. Scott, (1983). *Biosci. Rep.*, **3**, 141–151.
129. P.S. Mamont, S. Bey, and J. Koch-Weser, (1980). In *Polyamines in Biomedical Research* (Ed. J.M. Gaugas) pp. 147–165, John Wiley, New York.
130. J.J. Marr, (1983). *J. Cell. Biochem.*, **22**, 187–196.
131. J.J. Marr, (1984). In *Parasitic Diseases*, Vol. 2, *The Chemotherapy* (Ed. J.M. Mansfield), pp. 201–207, Marcel Dekker, New York.
132. J.J. Marr, and R.L. Berens, (1983). *Mol. Biochem. Parasitol.*, **7**, 339–356.
133. J.J. Marr, R.L. Berens, N.K. Cohn, D.J. Nelson, and R.S. Klein, (1984). *Antimicrob. Agents Chemother.*, **25**, 292–295.
134. P.D. Marsden, C.C. Cuba, and A.C. Barreto, (1984). *Trans. R. Soc. Trop. Med. Hyg.*, **78**, 701.
135. J. Mauel, (1984). *Mol. Biochem. Parasitol.*, **13**, 83–96.
136. J. Mauel, J. Schnyder, and M. Baggiolini, (1984). *Mol. Biochem. Parasitol.*, **13**, 97–110.
137. T.I. Mercado, and A. Colon-Whitt, (1982). *Antimicrob. Agents Chemother.*, **22**, 1051–1057.
138. S.R. Meshnick, (1984a). In *Parasitic Diseases*, Vol. 2, *The Chemotherapy* (Ed. J.M. Mansfield), pp. 165–169, Marcel Dekker, New York.
139. S.R. Meshnick, (1984b). *Pharmac. Ther.*, **25**, 239–254.
140. S.R. Meshnick, D. Brown, and G. Smith, (1984). *Antimicrob. Agents Chemother.*, **25**, 286–288.
141. M. Milhausen, R.G. Nelson, M. Parsons, G. Newport, K. Stuart, and N. Ayabian, (1983). *Mol. Biochem. Parasitol.*, **9**, 241–254.
142. D.H. Molyneux, and R.W. Ashford, (1983). *The Biology of Trypanosoma and Leishmania, Parasities of Man and Domestic Animals*, Taylor and Francis, London.

143. H.C. Nathan, G.J. Bacchi, S.H. Hutner, D. Rescigno, P.P. McCann, and A. Sjoerdsma, (1981). *Biochem. Pharmacol.*, **30**, 3010–3013.
144. H.C. Nathan, K.V.M. Soto, R. Moreira, L. Chunosoff, S.H. Hutner, and C.J. Bacchi, (1979) *J. Protozool.*, **26**, 657–660.
145. R.A. Neal, and S.L. Croft, (1984). *J. Antimicrob. Chemother.*, **14**, 463–475.
146. R.R.C. New, M.L. Chance, and S. Heath, (1981). *J. Antimicrob. Chemother.*, **8**, 371–381.
147. R.R.C. New, M.L. Chance, and S. Heath, (1983). *Biol. Cell.*, **47**, 59–64.
148. B.A. Newton, (1983). In *Chemotherapeutic Strategy* (Eds. D.I. Edwards and D.R. Hiscock), pp. 85–101, Macmillan, London.
149. L.L. Nolan, J.D. Berman, and L. Giri, (1984). *Biochem. Int.*, **9**, 207–218.
150. L.L. Nolan, and G.W. Kidder, (1980). *Antimicrob. Agents Chemother.*, **17**, 567–571.
151. K.K. Oduro, I.B.R. Bowman, R., and I.W. Flynn, (1980). *Expl Parasitol.*, **50**, 240–250.
152. P.O.J. Ogbunude, and C.O. Ikediobi, (1983). *Mol. Biochem. Parasitol.*, **9**, 279–287.
153. V.I. Okachi, A.M. Abaelu, and E.O. Akinrimisi, (1983). *Biochem. Int.*, **6**, 129–139.
154. F.R. Opperdoes, (1980). *Trans. R. Soc. Med. Hyg.*, **74**, 423–424.
155. F.R. Opperdoes, and P. Borst, (1977). *FEBS Lett.*, **80**, 366–374.
156. W.E. Ormerod, (1979). *Pharmac. Ther.*, **6**, 1–40.
157. M.A. Ouaissi, D. Afchain, A. Capron, and J.A. Grimaud, (1984). *Nature (Lond.)*, **308**, 380–382.
158. M. Parsons, R.G. Nelson, G. Newport, M. Milhausen, K. Stuart, and N. Agabian, (1983). *Mol. Biochem. Parasitol.*, **9**, 255–269.
159. R.A. Pascal, N.L. Trang, A. Cerami, and C. Walsh, (1983). *Biochemistry*, **22**, 171–178.
160. R.D. Pearson, A.A. Manian, D. Hall, J.L. Harcus, and E.L. Hewlett, (1984). *Antimicrob. Agents Chemother.*, **25**, 571–574.
161. R.D. Pearson, A.A. Manian, J.L. Harcus, D. Hall, and E.L. Hewlett, (1982). *Science*, **217**, 369–371.
162. A.E. Pegg, and J.K. Coward, (1981). In *Advances in Polyamine Research*, (Eds. C.M. Calderera, V. Zappia and U. Bachrach), Vol. 3, pp. 153–161, Raven Press, New York.
163. A.E. Pegg, K.-C. Tang, and J.K. Coward, (1982). *Biochemistry*, **21**, 5082–5089.
164. J.R. Piper, A.G. Lazeter, T.P. Johnston, and J.A. Montgomery, (1980). *J. Med. Chem.*, **23**, 1136–1139.
165. M.F. Pupkis, and G.H. Coombs, (1984). *J. Gen. Microbiol.*, **130**, 2375–2383.
166. M. Rabinovitch, J-P. Dedet, A. Ryter, R. Robneaux, G. Topper, and E. Brunet, (1982). *J. Expl Med.*, **155**, 415–431.
167. W. Raether, and H. Seidenath, (1984). *Zeit. Parasitkend.*, **70**, 135–138.
168. P. Rainey, C.E. Garrett, and D.V. Santi, (1983). *Biochem. Pharmacol.*, **32**, 749–752.
169. R. Ribiero-dos-Santos, A. Rassi, and F. Koberle, (1981). In *Antibiotics and Chemotherapy*, Vol. 30, *Antiparasitic Chemotherapy* (Ed. H. Schonfeld), pp. 115–134, Karger, Basel.
170. M.R. Rifkin, (1983). *J. Cell. Biochem.*, **23**, 57–70.
171. M.R. Rifkin, (1984). *Expl Parasitol.*, **58**, 81–93.
172. G.F. Riou, P. Belnat, and J. Bernard, (1980). *J. Biol. Chem.*, **255**, 5141–5144.
173. N. Robinson, K. Kaur, K. Emmett, D.M. Iovannisa, and B. Ullman, (1984). *J. Biol. Chem.*, **259**, 7637–7643.
174. G.W. Rogerson, and W.E. Gutteridge, (1980). *Int. J. Parasitol.*, **10**, 131–135.
175. L. Ruben, C. Egwuaga, and C.C. Patten, (1983). *Biochim. Biophys. Acta*, **758**, 104–113.
176. L. Ruben, J.E. Strickler, C. Egwuaga, and C.C. Patten, (1984). In *Molecular Biology of Host–Parasite Interactions* (Eds. N. Agabian and H. Eisen), pp. 267–278, Alan R. Liss, New York.
177. A. Ryter, J.P. Dedet, and M. Rabinovitch, (1983). *Exp. Parasitol.*, **55**, 233–242.
178. T.A. Salzman, A.M. Stella, E.A. Widerde Xifra, A.M. Del, I.C. Battle, R. Docampo, and A.O.M. Stoppani, (1982). *Comp. Biochem. Physiol.*, **72B**, 663–667.
179. L. Schnur, U. Bachrach, G. Bar-ad, M. Haran, M. Tashima, and J. Katzhendler, (1983). *Biochem. Pharmacol.*, **32**, 1729–1732.
180. J. Schottelius, (1982). *Zeit. Parasitkend*, **68**, 147–154.
181. T. Seebeck, and P. Gehr, (1983). *Mol. Biochem. Parasitol.*, **9**, 197–208.
182. A. Sjoerdsma, and P.J. Schechter, (1984). *Clin. Pharmacol. Ther.*, **35**, 287–300.
183. D. Snary, (1980). *Expl Parasitol.*, **49**, 68–77.
184. T. Spector, and T.E. Jones, (1982). *Biochem. Pharmacol.*, **31**, 3891–3897.

185. T. Spector, T.E. Jones, S.W. Laton, D.J. Nelson, R.L. Berens, and J.J. Marr, (1984). *Biochem. Pharmacol.*, **33**, 1611–1617.
186. E.A. Steck, (1981). *J. Protozool.*, **28**, 30–35.
187. L. Stevens, and E. Stevens, (1980). In *Polyamines in Biomedical Research* (Ed. J.M. Gaugas), pp. 167–183, John Wiley, New York.
188. H.H. Stibbs, (1984a). *J. Neurochem.*, **43**, 1253–1256.
189. H.H. Stibbs, (1984b). *J. Parasitol.*, **70**, 428–432.
190. H.H. Stibbs, and J.R. Seed, (1975). *Experientia*, **31**, 274–275.
191. M. Tanner, (1980). *Acta Trees, and W.E. Gutteridge, (1980). Int. J. Biochem.*, **11**, 117–120.
192. M.B. Taylor, H. Berghausen, P. Heyworth, N. Messnger, L. J. Rees, and W.E. Gutteridge, (1980). *Int. J. Biochem.*, **11**, 117–120.
193. K.J.I. Thorne, and J.M. Blackwell, (1983). *Adv. Parasitol.*, **22**, 43–151.
194. M. Turner, (1980). *Nature (Lond.)*, **284**, 13–14.
195. M. Turner, (1982). *Adv. Parasitol.*, **21**, 70–153.
196. M. Turner, (1985). *Br. Med. Bull.*, **41**, 137–143.
197. B. Ullman, (1984). *Pharm. Res.*, 194–203.
198. P. Ulrich, and A. Cerami, (1982). *J. Med. Chem.*, **25**, 654–657.
199. P. Ulrich, and A. Cerami, (1984). *J. Med. Chem.*, **27**, 35–40.
200. C. Van der Meer, J.A.M. Versluijs-Broers, and F.R. Opperdoes, (1979). *Expl Parasitol.*, **48**, 126–134.
201. C. Van der Meer, J.A.M. Versluijs-Broers, C.Th. Van Dun, T.S.G.A.M. van den Ingh, J. Nieuhlenhuijs, and L.D. Zwart, (1980). *Tropenmed. Parasitol.*, **31**, 275–282.
202. K. Vickerman, (1965). *Nature (Lond.)*, **208**, 762–766.
203. F. Villalta, and F. Kierszenbaum, (1983). *Biochem. Biophys. Acta*, **736**, 39–44.
204. F. Villalta, and F. Kierszenbaum, (1984). *Biochem. Biophys. Res. Commun.*, **119**, 228–235.
205. N. Visser, and F.R. Opperdoes, (1980). *Eur. J. Biochem.*, **103**, 623–632.
206. H.P. Voorheis, (1980). *Mol. Biochem. Parasitol.*, **1**, 177–186.
207. R.D. Walter, and F.R. Opperdoes, (1982). *Mol. Biochem. Parasitol.*, **6**, 287–295.
208. B.C. Walton, J. Harper, and R.A. Neal, (1983). *Am. J. Trop. Med. Hyg.*, **32**, 46–30.
209. C.C. Wang, (1982). *Trends in Pharm. Sci.*, **1982**, 354–356.
210. C.C. Wang, (1984). *J. Med. Chem.*, **27**, 1–9.
211. L. Weinrauch, R. Livshin, Z. Even-Paz, and J. El-On, (1983). *Arch. Dermatol. Res.*, **275**, 353–354.
212. J.S. Weldon, J.F. Munnell, W.L. Hanson, and C.R. Alving, (1983). *Zeit. Parasitkend.*, **69**, 415–424.
213. S.R. Welhausen, and J.M. Mansfield, (1980). *Cell. Immunol.*, **54**, 414–424.
214. G.T. Williams, (1984). *J. Cell. Biol.*, **99**, 79–82.
215. J. Williamson, (1979). *Pharmacol. Ther.*, **7**, 445–512.
216. J. Williamson, and D.J. McLaren, (1978). *Trans. R. Soc. Trop. Med. Hyg.*, **72**, 660–661.
217. J. Williamson, D.J. McLaren, and J.R. Brown, (1983). *Cell. Biol. Rep.*, **7**, 997–1005.
218. J. Williamson, and T.J. Scott-Finnigan, (1978). *Antimicrob. Agents Chemother.*, **13**, 735–744.
219. J. Williamson, J.T. Scott-Finnigan, M.A. Hardman, and J.R. Brown, (1981). *Nature (Lond.)*, **292**, 466–467.
220. M.S. Wysor, and J.P. Scovill, (1982). *Drugs Expl Clin. Res.*, **8**, 155–168.
221. M.S. Wysor, L.A. Zwelling, J.E. Sanders, and M.M. Grendu, (1982). *Science*, **217**, 454–456.
222. J.R. Young, E.N. Miller, R.O. Williams, and M.J. Turner, (1983). *Nature (Lond.)*, **306**, 196–198.
223. A. Zenian, and F. Kierszenbaum, (1983). *J. Parasitol.*, **69**, 660–665.
224. D. Zilberstein, and D.M. Dwyer, (1984). *Science*, **226**, 977–979.
225. O. Zumbutal, and H.G. Weder, (1982). *Biochem. Biophys. Res. Commun.*, **107**, 869–877.

# Index

UNIVERSITY OF RHODE ISLAND
3 1222 00922 9900